2023 SUDOKU

PUZZLE BOOK FOR ADULTS AND SENIORS

		2	6		9	1		
				7	4	6		
6	4		1					3
7	1					2		9
	8			4			5	
2		6					1	4
4					7		6	1
		7	4	1				
		1	8		5	4		

For permission contact:
permission@weaverfrogcreativepublication.com

THIS BOOK BELONGS TO:

CONTENTS

How to play Sudoku:

Rule 1: Use Numbers 1-9

Sudoku is played on a grid of 9 x 9 spaces. Within the rows and columns are 9 "squares" (made up of 3 x 3 spaces). Each row, column and square (9 spaces each) needs to be filled out with the numbers 1-9, without repeating any numbers within the row, column or square.

Rule 2: Don't Repeat Any Numbers

		5	3				2	1
9	3		7					
	2	1						
				8	1	3	7	
							1	4
				4	3	5	8	
	5	4						
3	9		8					
		8	9				5	7

As you can see, in the upper left square (circled), this square already has 5 out of the 9 spaces filled in. The only numbers missing from the square are 4, 6, 7 and 8. By seeing which numbers are missing from each square, row, or column, we can use process of elimination and deductive reasoning to decide which numbers need to go in each blank space.

Rule 3: Never Guess

It is a game of logic and reasoning. So, you should not guess. If you don't know what number to put in a certain space, keep scanning the other areas of the grid until you get a chance to place a number. But don't try to force anything. It rewards patience, insights, and recognition of patterns, not blind luck or guessing.

Rule 4: Use The Process of Elimination

One way to figure out which numbers can go in each space is to use Process of Elimination by checking to see which other numbers are already included within each square, since there can be no duplication of numbers 1-9 within each square or row or column.

Sudoku rules are relatively uncomplicated. But the game is infinitely varied with millions of possible number combinations and a wide range of levels of difficulty. But it is all based on the simple principles of using numbers 1-9, filling in the blank spaces based on deductive reasoning, and never repeating any numbers within each square, row or column.

SUDOKU
PUZZLES
Easy - {1 -200}

EASY - 01

7				9				6
		4				9		
2			5		3			7
		1				6		
3	7		6		2		5	4
8	6						7	1
	3			2			1	
		8	4		1	5		

EASY - 02

4							3	
		8			3	7		
7	9		8					
	2			5	6			7
	6	5				2	9	
8			2	3			6	
					4		1	2
		6	3			9		
	4							5

EASY - 03

	7	1				9	4	
		2	5	1	8	3		
1								7
	9			6			8	
		3		8		4		
		8	7		9	5		
4				2				8
		7				1		

EASY - 04

	6		9	4			7	
4						8		3
	5	3			7	6		
		8	3		6			9
1								5
5			4		9	7		
		9	6			4	5	
6		5						8
	4			9	8		3	

EASY - 05

					1			8
			6	5			1	
1		7					6	5
				9		1		2
3			5		6			4
4		2		8				
5	4					9		6
	1			4	3			
2			9					

EASY - 06

2		9			3			6
			8			7		
	1			9				3
9				4			6	
		4	2		6	5		
	5			3				4
8				5			2	
		1			2			
5			7			6		1

EASY - 07

	6	4			3		9	
		1					4	
	5	9	8					3
	1			6				9
		6	7	3	1			
	8			9				7
	2	5	1					6
		7					5	
	4	8			9		1	

EASY - 08

	2			9		1		
1	7				2			
9				4			3	2
	1				8			
	8		9		7	5	2	
	3				4			
6				8			5	1
8	4				1			
	5			6		8		

EASY - 09

1			3				9	8
9					5	1		
	8	5		9		4		
	9			5				6
		3	6	7	8	2		
8				2			5	
		8		3		9	4	
		9	4					1
6	3				9			2

EASY - 10

	4		3	1				6
	5	8			9	7	1	
		7						
5		2	7					
					2	1		4
						6		
	8	6	4			2	3	
3				8	5		7	

EASY - 11

		6	8					
4	1		2			9		6
	2	5				8		
7		2	1	9				
				7				
				8	5	2		7
		3				4	6	
2		9			4		3	8
					3	1		

EASY - 12

	9	5		8		4	7	
			7		5			
		3				9		
5	7			4			9	2
			2		9			
3								8
				6				
			8		3			
	4			2			6	

EASY - 13

9		7		1				5
			2		8			1
	6			4				
		1	8			3		
2		8				1		4
		4			5	9		
				8			2	
7			4		1			
8				3		4		6

EASY - 14

	3		7	2				
		8						1
		6			1	2	5	
		2		1				6
7			8	3	4			2
8				5		9		
	8	3	1			7		
2						3		
				7	8		2	

EASY - 15

		9					7	
5			9		3			
				2	8			9
				5		2	4	
	8	5				1	9	
	7	3		4				
6			4	7				
			3		2			1
	3					5		

EASY - 16

1		9	3					2
				9	4	5		
	3		6					9
	1					4		7
	7						9	
8		2					1	
3					1		5	
		5	9	7				
2					5	9		4

EASY - 17

			9		1			
		3				2		
1	7			8			4	9
	4	8		5		9	2	
				9				
		6				4		
2		9		7		1		5
	5						8	
3			5		8			2

EASY - 18

		2	5					
	7				4		3	
		8			3	5		6
	2	6		9				3
			2		6			
7				8		6	4	
5		7	4			1		
	6		7				9	
					9	4		

EASY - 19

		3		4		6		
			6		2			
4								7
3	8						1	4
5								3
	9		3	8	1		7	
6								1
				3				
8	1			9			5	2

EASY - 20

			4	5	1	9	3	
		6				2	4	
		5						8
		1			4			
	9		1		5		7	
			6			3		
8						6		
	5	7				1		
	6	3	9	2	7			

EASY - 21

				3				
			5		6			
2		3	1		9	7		6
	3	9				6	5	
		4				9		
		1				3		
8	9		4		7		6	3
3			9		2			7
	7			6			4	

EASY - 22

	7					1	9	
4			3	2	9			8
9					8			
	4	1	8		6		7	
	2						8	
	3		2		1	9	4	
			6					9
3			9	4	7			5
	9	4					2	

EASY - 23

	1							
	6				7	4	2	
		8			2		6	1
1				6		8		
	8			4				
9				1		5		
		5			6		1	8
	7				8	3	9	
	2							

EASY - 24

		2				3		
7	6						4	9
				4				
8		3				6		5
5	1						2	7
				8				
		8	7		6	1		
9	7			5			6	3
				1				

EASY - 25

	1	9			7		5	
	3							
4			1	9				
8	6			5			1	
				4			6	2
9	5			1			4	
7			6	3				
	8							
	9	1			8		3	

EASY - 26

2				1				3
1	9		3		6		4	5
	8						9	
		9	8	6	4	1		
		3				5		
	6						8	
4				7				2
	3		4		1		5	

EASY - 27

		8		1			6	
	2	7				1	8	
				6				
5			4				9	
	6		1	7	5		3	
	4				8			5
				4				
	1	6				4	2	
	5			9		7		

EASY - 28

		5				7	6	
				9	3			
	1	6	5		8			
			3	2			5	
	2					6		7
			8	1			9	
	7	1	4		9			
				3	1			
		3				2	1	

EASY - 29

						7		
	6	9		8	4			
			7				3	8
8	5					9		
3		4				2		7
		6					1	4
9	4				2			
			4	5		3	2	
		3						

EASY - 30

						4	5	
	3		5	4				
			8		6	7		
	8	9			2			3
	7		6		5			9
	4	5			3			8
			9		4	8		
	5		2	7				
						2	9	

EASY - 31

9	1	7				3		8
			7		8			5
5				1				6
	4		9		1		6	
		1		4		2		
	6		8		7		1	
3				5				1
4			1		3			
1		2				8	3	9

EASY - 32

	5					1		
7			4					
	9	1		7				8
		3			4			
	1		9		7	6		
		7			5			
	7	5		8				2
8			5					
	2					4		

EASY - 33

			1	9	2			
9		5				8		4
			5	1	6			
8			9		4			7
	6						3	
	1	4		6		7	2	
		7	2		9	4		

EASY - 34

					4			
	2			9		4	1	
	4	6	5	7		3		
2						5		
	5	1		4		2	3	
		7						6
		2		6	5	1	7	
	7	8		3			9	
			9					

EASY - 35

9				4	3			5
	6		9		5			3
				2			9	
		2				6		4
4				9				1
8		6				3		
	4			6				
5			2		9		1	
6			3	5				7

EASY - 36

	6	9		4		3	1	
5	1		6		3		8	4
7	2						3	6
			5	8	7			
		5	9	7	6	1		
	8			1			5	

EASY - 37

1			7					3
				1		4		
	4	8	5	3	6	1		
		5	3		8	2		6
	3	6				9	4	
8		1	6		4	3		
		2	4	5	7	6	3	
		7		6				
6					3			2

EASY - 38

5	8			3			7	1
			6		7			
		3		4		2		
3	2						5	4
6		7				8		2
				2				
			4		9			
		9		6		1		
4		8				5		7

EASY - 39

			1					2
3		7			8	5		
		8			4		9	
	9	2		4				
7			5			3		
	3	1		2				
		9			1		8	
8		3			9	7		
			4					3

EASY - 40

7	1		4				2	9
4			3	2		6		
9								
					6		3	
				1				
	6		9					
								6
		5		7	8			3
8	3				1		9	7

EASY - 41

		4		7			9	
	1				6			7
		2	3					
		1			2			8
		9	6		8	3		
8			4			9		
					4	6		
2			7				4	
	3			5		7		

EASY - 42

2	6				8			7
	3				7	8		6
	9					3		
		3	1					
8			7		4			9
					5	2		
		9					8	
6		7	9				1	
1			5				9	3

EASY - 43

1		4	7				6	8
3					8			
				6				3
	3		6	8	9			5
		1	4		7	6		
6			1	2	3		8	
2				1				
			2					6
7	5				6	2		1

EASY - 44

			3	2			6	
5	6			1			7	
		3			6	9		
	8				4	7		
3		1						
	4				9	5		
		5			2	1		
6	2			4			5	
			5	6			4	

EASY - 45

				1		6		
6								7
	5			6	9		4	3
3	7							8
		8					3	
1	4							9
	3			4	2		9	1
9								5
				9		2		

EASY - 46

4								3
		2	3			6		
	6	5	1			8	4	
				2		3	9	
			5	8	9			
	9	1		3				
	2	6			3	9	8	
		8			7	2		
5								1

EASY - 47

4			6			5		
5	8							6
			7		5		4	
2			8			1		
	4			3			7	
		5			9			8
	5		2		4			
7							1	9
		4			6			3

EASY - 48

9				1				2
			7	9	2			
7								4
		7				8		
	2			8			7	
8	4						2	5
		6	9		5	1		
4								7
2		5				9		6

EASY - 49

		6	9		7	5		
	5			3			9	
			6	2	5			
9								5
		8		4		1		
	2	4				9	3	
	7						8	
	6	1	7		3	2	5	
		9				6		

EASY - 50

					7	9	3	
			8	3	6			
4	7							
7			3	9			6	
	5						8	
	1			2	5			4
							9	6
			7	8	2			
	3	1	4					

EASY - 51

	1	5		4		7	6	
		2		1		4		
	3		1		4		7	
1	5			8			9	4
		7		3		8		
2			8		7			5
5		3				6		9

EASY - 52

	6	1		7		8	2	
		2	1		5	7		
				6				
			5	8	4			
		4				5		
8	3						7	9
	1						6	
		8				9		
9			4	1	2			8

EASY - 53

		5				2	8	9
		7	6					
			4		2			
	1			3				5
	8	4					7	6
	5			4				3
			7		5			
		1	9					
		9				3	1	7

EASY - 54

			4	9	3			
	7						1	
	3						4	
8		9				1		5
	1						9	
				5				
			7		6			
7								2
5	6		8	2	1		3	9

EASY - 55

	9		5			3		
				2		4		1
6	2				1			
		3		6				7
	7		2		3		5	
4				9		1		
			8				4	3
2		6		3				
		5			2		7	

EASY - 56

		9		2	7	6		
		4			3			
5							1	4
4	5			7				
2			3		6			7
				4			2	6
1	9							2
			9			4		
		7	5	1		9		

EASY - 57

				2	8		3	4
			6			7	9	
5								
	5		1		3	9		
		4				5		
		6	2		9		4	
								6
	1	7			6			
6	2		3	9				

EASY - 58

	1	5	2					3
				1		7		5
		3					6	
	2		5					
5		6	3		1	9		
	3		4					
		2					5	
				8		1		2
	7	1	9					6

EASY - 59

	1		4	8		7		
					7	3		
		9					4	
	3	7	8				2	
2					4			3
	8	5	1				9	
		6					5	
					8	1		
	9		7	6		2		

EASY - 60

								6
		8	3		7		5	
	5	9		1				
3			5					7
						6	8	9
9			8					4
	4	5		9				
		6	2		1		9	
								8

EASY - 61

	7	2		8			4	
						6		
5				6				2
2					7	3		5
	1				3			
9					1	8		4
7				5				3
						9		
	3	9		7			6	

EASY - 62

4	7						9	6
		9		4		8		
1			9	3	5			7
		3				6		
			7	5	6			
	4			9			7	
			3	8	2			
	8						2	
9								3

EASY - 63

		5		6		9	7	
8			2			5		
2	7		5					3
				7		2	9	
7			9		1			5
	5	3		4				
3					6		1	9
		7			9			6
	6	9		1		7		

EASY - 64

3	4						9	7
8		5				4		2
	5		2		6		3	
		8				2		
		1	8		5	9		
4	6			1			2	9
1	9	2				3	4	5
				9				

EASY - 65

				1				
5			7	3	6			2
	7						3	
		1	2		8	6		
7								4
	6		9		5		2	
		6	1		9	5		
	3						1	
8								9

EASY - 66

			3	1		6	7	
6		5						
4			5	9			1	
			7		3	9		2
3		9		5		1		7
7		2	9		1			
	7			3	5			9
						8		6
	8	3		4	9			

EASY - 67

				7				
6			5		8			9
2	8			9			1	6
			3		6			
1								3
5		4		2		6		7
		5				7		
3			1		7			8
				8				

EASY - 68

		2	7	4			5	
7	9						1	
					9			2
		3		9				6
9			4		7			1
6				3		8		
5			6					
	3						2	8
	8			2	3	1		

EASY - 69

	9		3				8	
	3			5	4			6
2					6		7	9
9		6		3		2		5
4	8		9					1
1			2	4			6	
	6				9		5	

EASY - 70

			5	2	9	3	1	
				1				
	6		7		8	2		
	2					4		
	8	6				1	2	
		1					8	
		5	9		2		4	
				6				
	1	4	8	7	5			

EASY - 71

				3	8		6	
5		6				7		
	3			9			2	
4			8		3			
7		1				3		9
			9		6			4
	8			6			7	
		2				8		3
	1		3	8				

EASY - 72

3						4		9
		2	1					
8			9		4		2	
		4	6		8	7	3	
	6	8	7		5	2		
	4		8		9			3
					6	5		
7		9						2

EASY - 73

						4	6	1
	9		8	2				
	3				7		9	
		8	5	1				
					9			2
		7	3	4				
	4				5		2	
	2		1	7				
						1	7	3

EASY - 74

	9			3		1		
			9	4	8			6
8					7			
	4	8					6	
9	5						8	7
	3					5	1	
			5					9
4			3	7	1			
		6		8			7	

EASY - 75

	1			4	6		9	
4			3	2				8
		3			7	4		
1		2					4	
3	4						7	6
	5					2		3
		4	6			9		
7				8	4			5
	3		9	5			8	

EASY - 76

4	5		2		3		8	9
2								5
		3		8		4		
		8				5		
			4	3	6			
	7	1		4		3	5	
			5		1			
9				7				2

EASY - 77

			3		7			
5		7				3		8
	1	9				7	5	
				7				
	8		9	4	1		7	
7		5	2		8	1		3
4		8				5		1
9	5						4	7

EASY - 78

		8	9	6	2	4		
	9						6	
		5	3		7	2		
	2	6		5		7	8	
	1		7		4		3	
			4		6			
5	7	4		9		8	1	6

EASY - 79

	6	9	2	7				
		2		9	1			7
			6				9	2
	9					3		8
8	7			3			2	9
3		6					4	
6	4				2			
5			1	4		8		
				6	7	2	5	

EASY - 80

					7	5	9	
7					4			
1			2	9				
8	4		1		6	3		
		1		2		8		
		2	5		3		1	7
				3	2			5
			9					4
	2	3	8					

EASY - 81

		6			7		1	
8				4	9			
			5					2
6	1		3		4	9		
	4			2			6	
		3	6		5		4	7
1					3			
			4	5				6
	2		1			8		

EASY - 82

2				9				8
		6		2		1		
		1				3		
			9		7			
9	6		2		4		1	7
7				6				5
5								6
	3						8	
	7		5	8	3		4	

EASY - 83

	5			3			1	
	1	2	9		8	7	3	
7				6				5
9	6		4	1	5		2	7
1			5		7			9
			3	9	1			
		4		2		3		

EASY - 84

	7			2		4	3	
		4		3				2
						5	7	6
		7			8		5	
6		5				1		8
	8		4			9		
8	4	6						
9				8		7		
	3	2		5			8	

EASY - 85

				3				
		8	4		9	5		
1		7				8		9
		5	7		3	9		
	7	1	6		8	2	5	
8	1						6	5
		4		2		3		
	3						2	

EASY - 86

					9			
	2	9		4	7		6	
		8	5	1		4	2	
8	5					2		
	9	2				5	1	
		3					8	6
	8	7		9	2	6		
	4		6	3		7	5	
			7					

EASY - 87

3								9
	1						2	
8			2		7			3
	9						4	
6								2
5	7	4				8	9	6
	8						1	
	6			4			5	
			1	5	9			

EASY - 88

			3			2		
				7				1
1		2			8		9	
	8			2	9	7	4	
2			4				1	
	1			3	7	5	2	
8		3			1		7	
				5				2
			8			6		

EASY - 89

	2	8		5		7	9	
		3				4		
5			9		4			3
6			4		9			5
				3				
8			1		6			9
	7		5		8		2	
		5		9		6		

EASY - 90

	1			3			5	
3	4						1	8
		6	1			2		
			6		5	7		
7				2				6
		1	3		7			
		2			9	3		
4	9						7	5
	7			8			2	

EASY - 91

				2				
	4	1	7		9	2		5
	5						7	
	7	4			3			8
	9		1					
	3	6			7			9
	2						5	
	1	7	5		4	8		3
				8				

EASY - 92

			3		6	9	2	
	5			2				
			4		7	5		
4			2					7
	1			7			3	
9					4			1
		8	7		1			
				8			6	
	4	1	6		2			

EASY - 93

			5			8	7	
				8	3			6
3	5			6		4		
		5						8
	3	4				5	1	
8						9		
		9		2			8	4
5			1	4				
	4	1			8			

EASY - 94

		5			3		8	
3			1	5				
			4					5
7				2		5	6	
	6		8		7		2	
	4	9		1				8
6					1			
				4	8			3
	9		6			1		

EASY - 95

			9		4			
9	4			5			8	7
1								5
7	9		6		1		5	2
		1				4		
			2		7			
	7	3				2	6	
			7		6			
6				3				8

EASY - 96

		3			4			
			6	9	8			
9			1	5		8	4	
6								4
5							3	6
2								5
8			4	3		7	6	
			2	8	7			
		2			6			

EASY - 97

1		6	2					8
	8					2		
			8			1		
7			6			9	1	
	4			7			8	
	1	2			5			6
		1			7			
		8					9	
2					9	3		7

EASY - 98

		2		4	3	5		
				5			1	
3		7			2	4		
					4		7	
5		9		1		3		8
	2		9					
		1	3			8		4
	4			9				
		3	4	2		1		

EASY - 99

	9			8		6		
2		8		4	7			
7						3		
	6			2	4			
	4						6	
			1	9			7	
		5						8
			5	7		1		2
		2		3			5	

EASY - 100

8	6		5	3			4	
		4	9					
			8			2		
							3	7
	5	7		9				
							2	4
			4			8		
		2	6					
6	8		2	1			5	

EASY - 101

	4		1		3	5		2
8	2	5	6				1	
	1				6	8		
			8		2			
		6	9				3	
	3				1	4	7	5
2		1	3		7		9	

EASY - 102

			7	4	8			
7		2	9		1	5		3
	9		3		5		7	
	5						2	
8	2	7				3	9	4
9								8
	3	9				7	4	
		1				9		
			5		3			

EASY - 103

		7		4	1		5	
	6	4		9		7		
3			7			9		
			9					5
				1	2	3		
			8					4
7			4			5		
	8	5		3		6		
		3		8	5		7	

EASY - 104

			3			9	1	
		5	6				8	
4		1						
2					3			
8		3	2	5		7		
6					7			
5		4						
		2	4				5	
			8			1	4	

EASY - 105

		1			3			
			4		2		1	
			1			9		5
5	1				7			4
8		9				6		
4	2				5			3
			3			7		6
			5		1		2	
		2			6			

EASY - 106

			6		7		1	
3	4				5	6		
						8		
	9			6				
8	5	1				9	6	4
				1			5	
		5						
		3	2				9	5
	2		9		1			

EASY - 107

		3				5		
2	1						6	7
9			2		4			3
			5		8			
	9						1	
1				4				8
		1				7		
4		2	7	3	9	8		1
7			1		5			6

EASY - 108

				4	1	8		2
	1	7			6			
							7	
8	9		1		3			
7			5	6		9	8	
4	6		9		7			
							2	
	2	6			5			
				9	2	3		5

EASY - 109

		9	1	5			6	
3		6	2					
							3	1
			7	3	2		4	6
4			6		5			9
6	7		4	9	1			
1	3							
					7	3		5
	5			2	9	6		

EASY - 110

					6			
8	5	1						
4			3			1		
9	2	4	5					
	6			7			8	
					4	6	5	9
		8			2			5
						9	6	4
			7					

EASY - 111

		1				3		
9			4	8	5			7
	2						6	
2				9				8
7		4				9		2
	9						5	
	8		5		3		7	
		2	7		1	5		
	7			4			1	

EASY - 112

9			4		8		1	
	6			3				
			1	6	5			
6		4				1		7
	3						5	
7		9				3		2
			5	7	6			
				4			6	
	1		9		3			5

EASY - 113

	2						9	
3		4			8			
					5	1	4	
2			3			4	7	
				4				
9			1			3	5	
					9	5	2	
6		8			4			
	7						8	

EASY - 114

5								6
		2		8		1		
	4	7		6		8	3	
3		9				5		2
1								9
			9	4	7			
			8	9	2			
2			3		4			7

EASY - 115

2		5				9		6
			5					
3		6			8	5		4
		4		1			7	
			6	7	2			
	9			5		6		
1		7	4			2		3
					7			
5		8				4		7

EASY - 116

		1		2		8		
			6	7	8			
7	2						3	6
				6				
	5	6	3		7	9	8	
	4						7	
	8	2		3		4	9	
5	7			9			6	2
9								8

EASY - 117

			6	9		4	1	
7		4						9
6				5				7
		8		7	6			
		6	2			8		
		1		4	9			
8				6				5
2		5						8
			5	8		7	9	

EASY - 118

8		6		3		1		2
	5	4		2		8	6	
		5				9		
2	6			8			3	5
	8						2	
			8	6	3			
6				5				1
	3			7			9	

EASY - 119

5				4		9	6	
	2				1	5		
	3		6					
	7		4	1			2	
8								7
	1			7	8		5	
					4		7	
		9	1				3	
	4	7		3				6

EASY - 120

	7						1	
		1				5		
9			6		1			3
5			2		9			8
7			8		3			5
6	8			1			9	2
2		5				7		1
				3				
	6			2			5	

EASY - 121

		1		7		3		
	6			9			7	
8			3		4			1
3		2				9		4
	5						8	
1								7
6		4	9		5	8		3
		8	6		7	1		
				8				

EASY - 122

		2		5	7	8	1	3
					6			
3			8				7	
		3	6	7			9	
		6				7		
	1			9	2	6		
	6				5			2
			1					
2	3	5	7	6		1		

EASY - 123

	5		6			8		
		3	7					5
8					2		3	
		1		7			6	2
			2		5			
3	7			8		4		
	2		4					8
9					3	5		
		8			9		2	

EASY - 124

9				6		1		3
		2			1	5		
8	6				5		4	
	3	7		2				
2			3		6			5
				4		2	3	
	2		4				9	7
		8	7			6		
7		3		9				8

EASY - 125

	2			9			7	
			6		8			1
		7			1		3	5
	5							
	4	3				6	8	
							5	
2	6		5			7		
3			8		9			
	9			2			4	

EASY - 126

		8		3		9		
3								8
	1						3	
		4	3		9	6		
		3	5	7	6	1		
			1		8			
5	3	1				8	7	4
8	7						5	6
	2						1	

EASY - 127

1			9					
9			3	7		8		
8	5	3			6	9		
	4	8	6			7		
		7			5	2	8	
		5	2			4	1	8
		6		3	1			9
					9			6

EASY - 128

	7				3	1		5
	6							
		5				9	2	6
7			4		5	8		
5			3		2			4
		3	8		6			2
8	5	7				2		
							9	
2		9	6				7	

EASY - 129

			5	1		8		
	7						2	
6		4			7	5		
		6		4				1
1			6		8			2
4				9		7		
		2	1			4		3
	5						7	
		3		8	6			

EASY - 130

						3	1	
	9	1	8			4		2
				9				5
1					9			
	4			3	6	9		
7					4			
				2				3
	7	5	9			8		4
						2	5	

EASY - 131

4	3		9			5	6	
	8		1					
9				6	4			7
	7	9						
					5	9		6
	5	3						
2				4	9			3
	9		2					
5	4		7			8	9	

EASY - 132

						4		6
		4			3		1	2
1		7					5	
7				6				
	6	5	4	2				
3				9				
4		3					9	
		6			2		7	5
						8		4

EASY - 133

9				7			5	
	4			3		8		
	1	3			6			
6					9	2		
				2				
		9	1					8
			9			3	2	
		7		5			8	
	5			6				4

EASY - 134

	5				7	2	3	
7	9						4	6
2			6					
9			3		1	7		
		1	7		5			8
					3			7
5	8						1	2
	3	7	5				6	

EASY - 135

		3			8		4	
4					6	3		
	8		4					7
1	4		5	6	3	8		
			1		9			
		9	8	2	4		5	1
5					2		7	
		1	6					3
	9		3			6		

EASY - 136

		2				8		
			2		9			
	4						3	
8	2	3	6		4	7	9	1
		4				6		
				1				
	5			3			8	
	3		5		1		4	
	7			4			6	

EASY - 137

1					8		4	9
	3					6		
7				9		5		
	2		4		7			
		7					8	
	9		5		3			
3				2		8		
	6					7		
8					5		1	4

EASY - 138

	9	7	5	1				2
6			8	2				9
			7	3		2		5
8		1	6		5	7		
			1	9		4		8
7			3	5				4
	3	4	2	7				1

EASY - 139

	3					5		
1					3			9
5					8			1
	7		1			3		
	5			4			1	
	6		5			2		
3					5			4
9					7			3
	4					6		

EASY - 140

			4	6	7			
	4	7	8		2	6	1	
				8				
	7	6		4		8	2	
3	5						9	7
	8						6	
7		1				3		8
5	6						7	2

EASY - 141

					2			4
	5		3			7		
	1			9		6		2
			5	4				
7		9				1		
			1	2				
	8			3		4		1
	2		8			5		
					6			8

EASY - 142

	7	1			4	3	2	
				8				
		3			6	4		
				4	1			2
9			3	6				
		2	6			7		
				7				
	4	6	5			9	3	

EASY - 143

				2		9	7	
	5					8		1
3		1	8	7		6		
					6	1		
			7		4			
		4	2					
		8		9	1	3		2
4		5					1	
	1	3		6				

EASY - 144

	2		6					5
	9	3	7	8	4	6		
	6			2			8	
					6		1	
2								3
	4		2					
	1			9			3	
		8	3	7	2	1	4	
9					8		5	

EASY - 145

5		9			8	1	6	
1		6						
			2		6		9	7
			8					2
8	4						7	6
7					3			
2	1		3		5			
						2		3
	3	7	1			6		5

EASY - 146

			5	7		4		
		2	4				1	
1					6			
	8			6		7		5
		4		5		3		
7		3		1			2	
			9					6
	1				3	8		
		6		4	7			

EASY - 147

	4	5				9	3	1
		9	4			7		
					1		4	5
	7					3		8
5			8		3			
	2					6		4
					8		6	3
		8	5			2		
	1	2				5	8	7

EASY - 148

	5			3	9	4	8	
8		9		6				5
		3	8					
				2	8			7
	1	8				5	2	
2			1	4				
					3	8		
3				8		9		6
	8	4	5	9			1	

EASY - 149

		5						3
6	9						7	
1			4	6	3			5
		6	2		4			
2							1	
		1	3		7			
3			5	7	1			9
7	5						2	
		4						7

EASY - 150

		7		1			5	
1			6		2			
		4	5			2		8
	5					8	9	
4				6				2
	3	9					7	
9		1			8	3		
			1		3			5
	4			2		9		

EASY - 151

						7		
	1				8		3	
4		7	3	9	6	1		
	6	8	7		5	2		
		1				3		
		4	9		2	6	8	
		5	6	4	9	8		7
	8		1				4	
		6						

EASY - 152

5		6		9		8		2
			5		8			
	4						3	
1	7			8			2	4
	2	3				5	8	
8			7		3			6
		7	1		5	3		
	5						4	

EASY - 153

		9	5			7		
					2	9		
5		1			8			2
		5	6		1			8
	9							
		3	2		9			7
1		2			5			6
					4	8		
		7	8			3		

EASY - 154

1		3		4				5
8						3	6	
		5					1	7
			3	5	4	6		
				8				
		8	1	7	6			
5	2					8		
	3	7						4
4				1		7		6

EASY - 155

	2				6			
	7		9	3			2	5
		4				6		
9				2			1	
	4		7		5		6	
	8			6				3
		7				8		
2	5			9	1		3	
			6				5	

EASY - 156

	1	7	8				5	
8			9			6		4
	6			7				8
			5		4		1	2
		4				8		
7	3		6		8			
6				4			8	
3		8			6			5
	4				9	3	6	

EASY - 157

	4			8	9		1	
			6	5				
9	6	5						
1	9				7			
5	8						2	7
			5				6	8
						3	9	4
				1	5			
	7		4	6			8	

EASY - 158

					5			
5		8			3		2	1
				6			7	
3		9				7	5	
			4		2			
	6	4				2		9
	9			4				
8	1		7			4		6
			3					

EASY - 159

4								8
	7		9		4			1
	8				6		3	
1		9		4				
						4	6	
8		2		7				
	9				7		4	
	1		2		8			7
5								3

EASY - 160

7	6				5		4	
						6		
5		3	4					
				8	1		9	6
8	7		6			4	5	
				4	3		8	2
2		4	9					
						2		
6	1				2		3	

EASY - 161

		3	5					
	1	4		6	9		3	
			8	3			7	2
	9					5		7
	2	7				9	6	
8		6					2	
9	7			1	3			
	3		4	8		7	1	
					7	3		

EASY - 162

	1			3			2	
5			8	9	1			3
			2	4	5			
			6	5	7			
2			1		9			5
	8						7	
	5						3	
7		4				9		6
1	6						5	4

EASY - 163

					8		3	
		5		9			7	
	8	6	1					
7	2				3			
		8		5		3		
			4				2	7
					9	1	6	
	5			4		9		
	1		2					

EASY - 164

			3		6			
				9				
		5	2	4	7	6		
2		7				3		5
				7				
6			8		2			1
	7			2			4	
1	6		5		8		7	2
	3						5	

EASY - 165

		8	6			3	2	
		9	7					5
				1				
	2							7
1	9		8		6		4	3
5							9	
				3				
7					1	9		
	5	3			9	8		

EASY - 166

		3		2	1		6	
1	5						9	
			9		3			5
6		5	8		9	7		
2								9
		7	1		2	5		6
7			6		8			
	6						7	8
	2		3	5		6		

EASY - 167

8		5					7	
					4		9	
		6	1	2		8		
						7	5	
2			6		3			8
	1	8						
		3		9	1	4		
	2		5					
	7					5		1

EASY - 168

					6		5	
4	6				2		9	
		9	3			7		
3	2			1		6		
			2		4			
		1		6			4	2
		7			1	9		
	8		4				7	5
	4		7					

EASY - 169

9	5		7	2				3
			3		6	9		5
					5		4	
	3					1		6
			8		3			
5		4					9	
	4		5					
3		5	9		1			
7				3	2		8	9

EASY - 170

					9	2		
		9		8		5		
6			5		2		1	
9			6					
		4		3		8		
					8			5
	8		7		5			2
		1		2		6		
		6	8					

EASY - 171

				6		1	3	
3	5	9					4	
2					4		5	
		5		4				
9			6	7	5			8
				9		4		
	2		4					9
	4					6	7	1
	9	7		5				

EASY - 172

		2	3				4	
3			7			2		
	6	9	2	1		5		7
						8	6	2
		1		7		4		
6	2	5						
1		3		6	7	9	2	
		6			1			4
	8				4	3		

EASY - 173

2	6					4		
	5		2				8	1
		1				3		
6			1				7	
		5	8		2		6	4
4			6				5	
		6				5		
	9		3				1	2
8	4					9		

EASY - 174

4								6
	6						4	
9	3	5				8	2	1
6	8		2		3		5	4
1								7
	9	3		1		2	6	
			8		6			
	7	6		5		9	1	

EASY - 175

	3		9	1		8		
	6				8		3	4
8					4			
	2	8	5		1			3
5								1
6			8		3	4	5	
			4					7
3	4		2				9	
		6		5	7		4	

EASY - 176

5				1			2	
						5		6
		2			4	8		
	9		1				3	8
	2		7					
	7		3				6	9
		7			1	3		
						4		2
4				9			8	

EASY - 177

	6						7	
3				1				6
			8	3	6			
1		4	5	6	2	8		9
	9						1	
		2				5		
			4		7			
7			9		1			5
		5		2		1		

EASY - 178

	9	4		6			5	
		1						3
			3		2			9
7							1	
		3	2		5	9		
	1							6
5			4		8			
8						7		
	4			3		5	6	

EASY - 179

		2		5	3	4	8	
				9	4			
4	3							
6		4		7			2	5
		3				6		
5	1			6		8		4
							7	8
			1	2				
	5	8	7	3		9		

EASY - 180

				4	6		5	
	4	3		9			2	
8						6		
4		9	1					
	6						7	
7		2	4					
1						5		
	9	4		2			8	
				3	4		6	

EASY - 181

	9			7	1	2		
4			9					
	2			8	6		3	
		7				6		3
5								8
9		2				5		
	4		7	3			8	
					8			2
		8	1	2			5	

EASY - 182

			1	5	3			
3								8
	4	5	8		2	6	1	
	5		2	3	9		8	
			7	8	6			
	3	4				7	2	
6	1			7			3	4
5								9

EASY - 183

	4						3	
				9				
	1	3		8		7	5	
4	2						6	1
	3	7				8	2	
1								5
			9		7			
				5				
	5	1	3		6	4	8	

EASY - 184

				6			9	
	6				8			
4	5	8						3
3				8	2	1		
	7				1		2	
5				4	3	6		
1	2	4						8
	9				7			
				2			3	

EASY - 185

	8				5		1	
5	1			8	7		3	4
		7			3	8		
6	7	1						
	9			1			4	
						9	6	1
		4	8			1		
1	6		9	7			8	5
	2		5				7	

EASY - 186

6			4					
	5				1	4		
	4	8					7	
1	3				2	6		
4				5				8
		6	9				3	7
	7					8	4	
		2	8				1	
					3			2

EASY - 187

4			9	6				1
		9		8	2	6		
	5		4				3	
	8					4		6
3	9						7	5
2		4					9	
	3				7		4	
		7	1	4		3		
9				3	8			7

EASY - 188

6	9					8		
				7	8			1
	7				5			
	3		6		7		5	
4			5				1	6
	5		3		1		4	
	2				6			
				1	4			2
7	1					4		

EASY - 189

		5		8		1		
	4			5			6	
			3	4	9			
6			8		5			3
3			7		4			5
		4		1		6		
	6	1		7		3	8	
	8						4	

EASY - 190

2					3			
						3	2	5
8	9			5	6			7
		5	8	1			9	
	4							
		6	4	3			5	
6	8			9	5			4
						5	6	2
5					1			

EASY - 191

		5	8		3	6		
		4		2		9		
8				5				1
5			9		2			3
		3		4		5		
		8		9		3		
2			6		1			8
1								6

EASY - 192

1	9		5			7	8	2
3		8	7					1
5				1	8		9	
		1					7	5
		3				6		
2	5					8		
	3		9	4				6
7					6	9		8
9	1	6			2		5	7

EASY - 193

	6						9	
4	8	3				6	2	7
7								3
		6				4		
	1		2		7		6	
			3		1			
		8	6		4	5		
	9		5	2	8		3	

EASY - 194

			2	7	4			
3				6				2
	5			3			6	
	2						8	
				1				
	4	5				3	2	
	7	3	1		8	6	4	
	8						7	
		2	6		7	9		

EASY - 195

			1		7			
9	8						7	6
	5			3			2	
	7			2			4	
1								9
		8	4		3	7		
		4		1		3		
				4				
			7		6			

EASY - 196

				4		5	7	
	5							8
4				7	6		9	
		8			4			
			7		1		2	
		1			9			
2				3	7		6	
	3							5
				1		2	4	

EASY - 197

			2	3	8			
7								8
			7	9	4			
				4				
1	5			8			7	4
4		7				6		1
5	6						8	9
8		1				7		3
		4		6		1		

EASY - 198

		6				1		
			3	7				6
	9		5		1		7	
1					7			8
	5	7		2				
4					9			5
	1		8		5		4	
			7	3				1
		4				3		

EASY - 199

4			2			1		9
	8			3			7	
6		5				4		
			9		5			7
	2						4	
7			8		4			
		8				7		1
	1			9			5	
5		2			7			8

EASY - 200

	9	8			3		7	
6				1	5			4
			9					6
2	3			9		4		
	8		6		4		5	
		4		2			3	7
8					9			
4			7	8				9
	7		3			5	8	

SUDOKU
PUZZLES
INTERMEDIATE - {1 -300}

INTERMEDIATE - 01

4			1	6				
8	1	7			3			
5					4			
	4				7	2		
			2		9			
		9	3				7	
			8					1
			5			8	3	4
				3	2			5

INTERMEDIATE - 02

	9	2			1	6	7	
		5	6				4	2
4				3				
9	5	3			8			1
1				7				
		1	8				9	3
	4	6			5	1	8	

INTERMEDIATE - 03

3								
		6	1	8		7		9
	8					1	6	
				7				6
5	4	8						
				1				2
	5					6	8	
		3	8	6		2		7
8								

INTERMEDIATE - 04

	4			9			7	
2	3				5	9	4	8
	8				7			
	9	8	3		2			
3								6
			9		6	3	8	
			7				5	
9	7	3	6				2	1
	5			2			3	

INTERMEDIATE - 05

8								9
				4				
	4		1		8		2	
	1	8		5		3	6	
		9	2		4	5		
		4		6		9		
7		6	9		5	8		3
			8		3			

INTERMEDIATE - 06

			1	7			3	
7					2			
		6	9	4		2		
	1					7		4
4		8				9		2
5		9					6	
		3		6	5	8		
			4					3
	5			3	9			

INTERMEDIATE - 07

	8	1	4		7	9	5	
	4						6	
				9				
	2		1		9		7	
			5	8	2			
3								9
1								8
		5		2		3		
6		2				5		1

INTERMEDIATE - 08

	8		4			3		
		4						2
3		9				4	6	
			6	1	9			7
			3		5			
1			7	4	2			
	2	3				9		8
7						5		
		5			6		2	

INTERMEDIATE - 09

			4		1			
	8						3	
	6		2	5	3		4	
6		8		3		5		4
				6				
		7		1		8		
8	5			2			6	3
7		6		4		1		5

INTERMEDIATE - 10

7		5			9			
			2		5		9	
		4	8				6	
						6	8	
4		6	7		2	1		5
	1	8						
	4				8	9		
	3		6		7			
			1			7		6

INTERMEDIATE - 11

1			7					5
7					8	4	1	2
4	2		1					
	1					3		4
				7				
6		2					5	
					7		4	9
2	4	3	9					8
8					5			1

INTERMEDIATE - 12

		8						
5	1				6			
	9		7			2		5
9	7		2		8	6		
				4				
		3	6		7		5	2
7		1			4		3	
			3				1	9
						8		

INTERMEDIATE - 13

		9		4		8		
			9		5			
			8		6			
6	4						9	7
		1				4		
3	8		4		9		5	6
	9	8	3		2	7	6	
1								4
			7	1	4			

INTERMEDIATE - 14

	1				4			
8				5			4	
3			1		2		8	5
		5				4		2
	6						7	
		8				6		3
6			5		9		1	4
2				4			3	
	8				6			

INTERMEDIATE - 15

			1		8			
		2		7		5		
4		1	5		2	6		7
6			7	5	1			2
8								6
1			8		4			9
	4	5				9	2	
9								4

INTERMEDIATE - 16

	4	7		6	3			
	1		9					
9		3						4
1					9	6		2
			4	2	5	8	1	
7					6	4		3
4		1						6
	6		7					
	7	9		4	1			

INTERMEDIATE - 17

4				6				3
		3	9		1	5		
		1				7		
			4	5	8			
	4						2	
	1	9				4	7	
			8		5			
	6						4	
9				4				8

INTERMEDIATE - 18

		5			4		9	
2			7	5				3
			3				4	
1	5		6				7	
			5		9			
	3				7		8	5
	1				8			
4				7	5			6
	2		9			1		

INTERMEDIATE - 19

				6		3		
					4		9	1
1		9	8				7	
			7				1	6
				9				
7	2				3			
	7				9	2		4
4	8		6					
		5		1				

INTERMEDIATE - 20

		2		9	3			
	9		5				2	
				6	1			3
1		6					9	
4		9				2		7
	2					6		4
8			6	1				
	4				7		3	
			3	5		8		

INTERMEDIATE - 21

	2	7	4				1	
				2			5	
					9			3
		4			3		8	2
2								1
6	1		2			3		
7			3					
	9			4				
	3				1	8	4	

INTERMEDIATE - 22

7		2			3	4		
				4				
			5	9			2	3
9	4		8			7		
	6						5	
		7			4		1	9
6	3			2	1			
				6				
		9	4			6		7

INTERMEDIATE - 23

		7		5		2		
4								1
	5			4			6	
			9		6			
			4	8	2			
6				3				8
		6				1		
	7	2				5	8	
	3	5		2		9	4	

INTERMEDIATE - 24

	5		1		7		8	
	7			9			6	
9		4				7		3
		3				1		
		2	3		1	5		
			7		6			
8			6		5			1
6		5		1		8		4

INTERMEDIATE - 25

1		6	4		2			8
				6		1		
	4	3			7			
4							6	
	5	9	8	3			4	
7							1	
	6	7			5			
				7		6		
3		8	6		4			5

INTERMEDIATE - 26

			3		1			
	8	3				1	5	
1	2		5		9		6	7
3				8				9
5				7				2
	6						9	
			6	9	7			
	5						2	

INTERMEDIATE - 27

	5	1				2	7	
3		6				4		
					6		1	3
				9				7
	8	7	2				5	
				3				1
					4		3	2
2		8				1		
	4	3				9	6	

INTERMEDIATE - 28

8								7
	4		1		3		2	
	2	1				6	4	
	8			2			6	
1								8
		6	8		4	3		
	5			3			1	
	1		2		5		9	
				9				

INTERMEDIATE - 29

			2				5	
		9		3	5	1		4
			9					8
2			7		4			
	7	8		9		5		
6			5		8			
			1					2
		2		8	3	7		6
			6				8	

INTERMEDIATE - 30

		6	9	7		1		
								3
	1			6		5	7	
2		9						1
			4				2	
7		8						6
	6			2		7	9	
								8
		3	7	5		6		

INTERMEDIATE - 31

		8	2	4		6		
			7					
4					6			7
		4		7			5	3
2			5		3			4
5	7			9		2		
9			6					2
					8			
		6		3	1	7		

INTERMEDIATE - 32

3		4		6	5		2	
2				7				
	1							
	8		9	5		4		6
4	2						8	7
1		5		8	4		9	
							6	
				9				3
	4		5	3		2		1

INTERMEDIATE - 33

	6					8		2
				2	5		6	1
		8			6	9		
7					2			
		3		8		6		
			9					3
		4	7			3		
9	3		4	6				
5		7					4	

INTERMEDIATE - 34

		9	3			8		
		3			4			
5				8			1	3
	3		5		2			8
		7				6		
2			6		7		9	
7	2			4				5
			2			7		
		6			1	2		

INTERMEDIATE - 35

	4						7	
		2				9		
		3				8		
2				1				6
3	5		4		7		1	9
		5	1		3	4		
4			2		9			5
	7		5		6		3	

INTERMEDIATE - 36

	9		4					7
	2			7				
					2	1		5
5	8					2		1
2			6				4	
7	4					8		6
					7	6		2
	1			9				
	7		1					3

INTERMEDIATE - 37

	4				6			
	5	9	2				6	
7		6	1					5
	7		8		2	6		
		1	5		9		7	
4					8	3		6
	3				1	5	2	
			4				1	

INTERMEDIATE - 38

	1						8	
2				9				4
			5		3			
		1				6		
			3		1			
	7	6	2		8	5	1	
	8	9				4	7	
	5		8		7		3	

INTERMEDIATE - 39

			6		2			
1								9
3	7			1			6	8
	5						8	
2				8				6
		9	7		5	4		
			2		9			
9	3			6			4	1
				4				

INTERMEDIATE - 40

	2					6		
		4	9					
			3		2	1	9	7
		3	6				5	
1				2				6
	9				7	8		
7	5	2	8		3			
					4	3		
		8					2	

INTERMEDIATE - 41

7		6	5		8	9		3
2		1				5		7
5		7	2		9	8		1
		2		5		7		
			4		7			
			7		5			
			9	2	6			
9				8				6

INTERMEDIATE - 42

		2				8		
	1		8		7		4	
		4		1		3		
				6				
			1		2			
4	6						9	1
7		1	6		4	5		9
		3				6		
			3	9	8			

INTERMEDIATE - 43

4	1	6						
5	8			7			4	
			1		6			
				1		7	8	6
	9				2			
				3		9	2	5
			7		1			
1	3			2			9	
8	2	7						

INTERMEDIATE - 44

			1	8	6			
	6		5		7		3	
		7				5		
		3				6		
		4	2		8	9		
8		1	6		3	7		5
1								7
				7				
	3			4			9	

INTERMEDIATE - 45

	4	6	7		8	1	9	
		1	6		4	8		
				6				
2			1	8	5			6
	6						3	
		5				2		
8				1				9
1		4				7		8

INTERMEDIATE - 46

9			6		5			7
		3	1		4	8		
			2		8			
		6		8		1		
	9		3	5	7		4	
	4						9	
3		9				4		1
	6	7				3	2	

INTERMEDIATE - 47

	5							3
						2	7	6
		4	2		1	9		
	7			2			9	
8				4			5	
	3			9			1	
		2	9		6	7		
						5	6	4
	1							9

INTERMEDIATE - 48

			5		3			
3								2
	2	8		4		9	3	
		2	1		6	3		
	8						6	
4								9
	3						9	
		1		5		6		
		9	8	3	1	4		

INTERMEDIATE - 49

		2		7		9		
	7		4				5	
4				8	3			
						6		5
1			6		9			2
7		5						
			2	4				6
	4				1		8	
		6		9		3		

INTERMEDIATE - 50

					2			
		2	8					
		6	3	9		5	2	
9				2		3	1	
		7	1		6	9		
	2	8		3				7
	1	9		8	5	4		
					7	2		
			6					

INTERMEDIATE - 51

8	2	4			9			
		3	8	1				
9					4			
	4		6			2		
1			9		8			3
		2			5		6	
			7					8
				9	2	6		
			4			5	1	2

INTERMEDIATE - 52

		2	3		7	5		
	5			8			7	
	8		6	9	2		1	
	1	4				6	8	
				1				
	2	5	9		3	1	6	
				6				
9	3						5	7

INTERMEDIATE - 53

	4						8	
3		5		4		6		9
8	7						1	3
	5			9			3	
			4		6			
4		8				2		7
	6		1		4		9	
5			9	8	3			6
		4		6		3		

INTERMEDIATE - 54

		6		4	3		8	1
	8							7
					6	5		
8		4	1			9		
	6			3			1	
9		1	6			4		
					1	7		
	7							3
		8		2	7		9	4

INTERMEDIATE - 55

	7	1					5	6
5	3		1					4
2								
				5	7	4		
	6					3		
				8	6	1		
6								
9	4		8					7
	8	5					1	3

INTERMEDIATE - 56

		8	9					1
9	6			8		5		
2				5		7		8
1					9			
				2				
			4					2
4		6		1				9
		1		3			7	6
5					6	4		

INTERMEDIATE - 57

	3					2		
2			3	1				7
			6	8		3		
		4		2		9	7	
	7						4	
	1	2		4		8		
		1		6	8			
4				3	1			9
		7					6	

INTERMEDIATE - 58

				2		5	8	
6	5	2			1		4	
	7							3
		3			2			4
			1		4			
4			3			1		
5							6	
	3		4			8	5	7
	4	7		6				

INTERMEDIATE - 59

	4				9	7		
						1	3	8
2				3		4		
			2		5		7	1
5					8			
			6		4		9	3
9				4		3		
						9	6	5
	6				3	2		

INTERMEDIATE - 60

	2			8		6		
8							7	
					4	8		2
	7		9			4	6	
	4		7		1		2	
	5	9			8		1	
9		5	8					
	1							6
		2		7			4	

INTERMEDIATE - 61

4						9		
		5		7	8			1
9					2	3		8
8		6	3		4			
		4				8		
			2		6	5		9
5		9	1					3
3			8	4		1		
		8						4

INTERMEDIATE - 62

	7			2	1			
6	1						8	
			4		6			
3	6		2			5		7
		7				9	1	
9	2		5			6		8
			1		5			
1	9						6	
	5			6	7			

INTERMEDIATE - 63

						3	5	
1		5	2					
3				8	6		4	
		2		5			6	
		6	3		4	8		
	3			7		9		
	5		4	6				8
					7	5		1
	8	9						

INTERMEDIATE - 64

8				1		5		7
			3					
7				9	4			
1		8		4		7	9	
		5				4		
6		7		5		1	3	
2				7	8			
			4					
4				6		8		3

INTERMEDIATE - 65

6					8	2		
	7		9					
2				5	4	9		
		8		7				1
7	1		3				2	
		6		9				7
1				4	6	7		
	6		1					
5					7	1		

INTERMEDIATE - 66

	1							9
5				2	3		6	
							5	7
3		1			6			
	7	5		1			9	
6		2			8			
							3	6
1				6	4		8	
	3							4

INTERMEDIATE - 67

							9	
	1			7		2	5	
			3	2	1	7		
	3	6			8			2
1					2			
	8	5			3			4
			1	3	7	9		
	5			6		4	3	
							6	

INTERMEDIATE - 68

2						8		
		7			2		5	
		3		9		7		4
3	9			8				
				4		5		
6	4			3				
		6		7		4		2
		9			3		8	
5						9		

INTERMEDIATE - 69

				8		6		
1						7		
2		8		1			5	3
6	5				8	1		
			6		9			
		4	3				6	5
3	2			9		8		7
		6						1
		7		3				

INTERMEDIATE - 70

			4			7		
9					6		1	5
		2			3			
	5						8	
6			2	8	7			
	7						6	
		4			5			
1					4		2	6
			3			9		

INTERMEDIATE - 71

7				2				5
	1		3		7		8	
	6	8				9	3	
				5				
6	8		1		9		5	2
9								3
1								8
			6	7	4			
	5						6	

INTERMEDIATE - 72

	1					9	6	
7		2						
	3	9			5		8	
			5				7	6
8		1	7		9			
			1				3	9
	7	6			4		2	
4		5						
	8					7	4	

INTERMEDIATE - 73

		6		5	4	2		
	5				1			
8	4	2						
		3			9			5
6				7				2
9			6			4		
						8	2	3
			2				9	
		7	3	9		5		

INTERMEDIATE - 74

	6	9				2	3	
8								1
5				9				8
			9		8			
6		8	4		3	1		5
				6				
		2				5		
				3				
	9	5	6		1	7	2	

INTERMEDIATE - 75

1	7						2	6
4				1		8		
		9	6					3
			9	6				
	1					4	9	
			4	3				
		2	5					8
5				8		3		
7	8						5	4

INTERMEDIATE - 76

	6						7	
5			4		6			9
			1	2	5			
4		6		1		2		8
	1						6	
2		4				3		7
	8		2	3	9		5	

INTERMEDIATE - 77

4				2		5		
1			7		3		6	8
	9							
			9				4	
6		7		3	5		8	
			8				1	
	6							
5			3		9		7	2
8				4		1		

INTERMEDIATE - 78

2			7		1			3
	1	5		3		2	7	
				5				
		3	1	8	7	5		
		2	6		3	4		
		6				3		
			9		2			
4		7				9		1

INTERMEDIATE - 79

					4			
7				5				6
9	4	1	6	2				
	5			4		6		7
8						2		1
	1			3		8		4
1	8	2	3	6				
4				1				3
					8			

INTERMEDIATE - 80

	4					3	1	
		9		7		2		
					2		9	6
	5		7	4				
	7				9			
	8		3	1				
					7		5	8
		4		8		1		
	9					6	7	

INTERMEDIATE - 81

	9		6	8			4	
6			9			2		
		4		7				6
7				1			5	
	3					1		
8				6			3	
		5		3				8
1			4			5		
	8		2	5			7	

INTERMEDIATE - 82

		7	5		6	3		
2	1		9		3		7	6
			8		4			
	5						9	
6	7						5	4
3								7
				6				
9				4				8

INTERMEDIATE - 83

5	1			2	6		7	
				3				1
		6	5					
		2	1			5		
7		4		6		1		3
		1			4	2		
					2	8		
6				5				
	8		6	9			3	2

INTERMEDIATE - 84

		9	1		7	2		
		8	9	3	2	7		
			7	9	5			
		1				6		
	5						7	
	8						1	
2		6	8		1	3		7
	3	7				4	6	

INTERMEDIATE - 85

		9				3		
6	1		5					
2	5				9			6
					1	7		4
			8	4		2	1	
					7	6		5
9	7				6			1
3	2		1					
		1				5		

INTERMEDIATE - 86

					5		4	
7		6	4		3			
		2	1	9		6	3	
2	6					4	1	
		7				5		
	1	5					8	9
	2	1		3	4	8		
			7		2	9		4
	7		9					

INTERMEDIATE - 87

5		9						
3	4		1					
				2		8	5	
	9	8			3			7
	1			4			2	
4			8			1	6	
	2	1		7				
					4		8	1
						7		5

INTERMEDIATE - 88

					7	4		
	5				4		7	
	7	4	2			6		5
				2	8			6
			9		6			
6			4	1				
4		7			1	9	8	
	1		6				3	
		9	7					

INTERMEDIATE - 89

1				5				3
		9				7		
7			4		1			5
2		6	5		7	9		1
9								8
		4	6		2	3		
6		5				4		2
4	7						3	6

INTERMEDIATE - 90

		8		2		4		
	6		9		5		2	
			5		4			
	2						8	
			2	3	7			
	4	5	8		9	3	1	
3		2		7		9		5

INTERMEDIATE - 91

5		6					9	
	8		1					3
1	2			9				
	4		8	3			2	
			6		5			
	6			1	7		4	
				7			3	2
9					4		1	
	5					9		4

INTERMEDIATE - 92

5			4	2				
2						4	5	
	8	3	6					
				7		8		
6	2						1	7
		4		5				
					8	7	3	
	7	6						9
				9	7			5

INTERMEDIATE - 93

6		5		3				
	9				1	3	4	
					2	5		
	1		8	9			5	
		8		1	3			
	6		2	4			3	
					9	8		
	5				8	4	7	
2		4		5				

INTERMEDIATE - 94

9		8				2		7
	4						9	
	6	1				8	4	
				1				
	1		2		9		7	
		4	6	5	7	9		
6		5	1		4	7		2
4	2		3		5		8	6

INTERMEDIATE - 95

	9		4				3	
4		8				2		
	3	6	5					9
	4	5						
			6		5			
						4	9	
7					1	3	5	
		3				1		6
	6				7		2	

INTERMEDIATE - 96

		2				1		3
		3					8	
			6		4			
2				5	9		1	
7			4		6	3	2	
6				8	3		7	
			8		5			
		7					3	
		5				2		1

INTERMEDIATE - 97

7		4		8				
		3				4		
	1			7	3	8		
		9			4		7	
4				9			1	
		1			5		3	
	3			4	8	6		
		5				1		
9		6		5				

INTERMEDIATE - 98

				5				
			3		4			
7								5
8				4				2
	2	4				6	7	
	3			7			9	
		2		9		1		
		7	6		2	4		
	6		5		8		3	

INTERMEDIATE - 99

4			3	9				6
			5			2		1
					8		7	
7		3						4
1				4		9		
6		2						5
					5		9	
			7			1		2
9			2	6				8

INTERMEDIATE - 100

		8	4		2	1		
			9		5			
5								6
	5						7	
6	9			1			5	4
7		4				9		1
2			7		8			9
		3		9		8		

INTERMEDIATE - 101

1				2				
	2	8		6			7	
4			9					5
	4				6			
7	6						3	8
			4				6	
9					2			6
	8			3		1	2	
				8				7

INTERMEDIATE - 102

6				2	8			5
		2	5					
			9		3		4	
1		8				5	2	
2				5				3
	7	3				8		4
	3		7		2			
					6	2		
4			3	8				6

INTERMEDIATE - 103

		3	4	1				
7	8			5	9		1	
5		4			8			9
8	3							
				6				
							6	5
2			1			7		6
	9		2	8			5	1
				4	7	2		

INTERMEDIATE - 104

		7		2		1		
6		8		3		4		9
			9		3			
				7				
7	1		2		8		4	5
	7	6				2	1	
1		3				7		4
	4			8			9	

INTERMEDIATE - 105

				5				
	8		2		7		3	
	2						1	
			9		6			
3			1		2			5
	1	7		3		6	4	
6								8
			8		4			
7	4						6	2

INTERMEDIATE - 106

			2					
			7	1	6		9	5
9		1						
	8				1	5		
				3	7	6	1	
	5				2	9		
4		7						
			9	2	8		7	1
			6					

INTERMEDIATE - 107

3				1				
	8		6			3		
	4		5	3			8	
			4	5				6
6	7		3		2		1	8
4				8	6			
	1			6	5		9	
		8			1		2	
				7				1

INTERMEDIATE - 108

		9	4		2			7
			1			9		
1					7			4
6		1			4			
				7	1			3
8		5			6			
2					3			8
			8			7		
		3	6		9			1

INTERMEDIATE - 109

3			9		1			8
				2		3		
	2		4	7				
7						1		5
	4	1				7	8	
8		5						9
				5	2		1	
		6		8				
5			3		4			6

INTERMEDIATE - 110

9		1				6		
				6		8		
	2		4		1			9
1	8							
7		6	5		9	4		8
							9	7
4			1		5		8	
		5		8				
		7				3		5

INTERMEDIATE - 111

		6				3		
			1		8			
	7						1	
			9	4	5			
		8		3		4		
5			8		1			2
				6				
	9		3		2		6	
	2	3				7	8	

INTERMEDIATE - 112

				9				
			5			1	8	2
5					6		3	
4		8		1				3
6			2			8		
1		5		4				6
9					3		1	
			1			3	6	9
				5				

INTERMEDIATE - 113

1	6							7
		2	4	1		5		
5					8		6	
					9			
2	5		7	4	3		1	9
			1					
	7		9					6
		1		5	7	9		
8							7	1

INTERMEDIATE - 114

				4				
			1		8			
4			7		2			1
2	7						5	9
		5	4	8	9	6		
		2	9		4	5		
9	6		3		5		1	7
		3				2		

INTERMEDIATE - 115

		6	7		3	4		1
	5		6				2	7
2	3		5					
5		4		1		3		9
					4		6	2
3	4				5		7	
7		5	9		6	1		

INTERMEDIATE - 116

		6		5		7		
	4		3	2	1		9	
	2						8	
	8			3			7	
		1	4		6	9		
4			2		7			8
	6						5	
		8				1		
2								9

INTERMEDIATE - 117

8	3		5		4			9
				1			8	
1	5		2					
		5			3			
3		1				6		7
			9			8		
					7		4	1
	4			9				
9			4		5		7	6

INTERMEDIATE - 118

2			9		5			1
		6		3		5		
		3				4		
4	9	8	1		6	7	3	5
3								6
			3	8	7			
6								9
	4			5			7	
			2	4	9			

INTERMEDIATE - 119

8	3		7					1
		6		9		7		
					2			
5			8					4
6				4			3	
7			2					9
					3			
		2		7		4		
4	9		1					3

INTERMEDIATE - 120

	1		3		7		8	
						2		
8	9		1	6			7	
	3		2					4
9				1				8
5					9		1	
	5			9	1		4	6
		9						
	4		7		6		5	

INTERMEDIATE - 121

			5	3	2	8		
	4							5
	5	7			1			
					6			2
	8	6				1	5	
5			3					
			6			7	2	
6							8	
		5	7	8	3			

INTERMEDIATE - 122

				8				
2			6		4			5
		6	5	3	2	7		
9	7	5				2	8	6
			7		9			
	3			2			7	
		1				9		
		7				6		
3	2						1	7

INTERMEDIATE - 123

1		3	6				7	9
9						6		
	6		4					8
				2		7		5
			1	8	6			
2		9		7				
4					9		2	
		5						6
8	9				5	3		4

INTERMEDIATE - 124

	1	2	3			6		
			2	6				5
		7			9			
3				8	6			
	4						1	
			4	7				2
			7			9		
5				2	1			
		1			8	2	7	

INTERMEDIATE - 125

7	6			5			8	1
8		1				4		2
		3		6		7		
			1		7			
1								6
	4						9	
2				3				5
			5		4			
	7						2	

INTERMEDIATE - 126

	7				1		4	9
					4	3		
			6	3	8			5
	3	6						
8	9	4		6	3			
	1	5						
			1	4	6			2
					7	9		
	6				5		1	3

INTERMEDIATE - 127

								6
3				7			9	
2	4				9		3	
6					3			9
	2	5	6				7	
4					2			8
8	3				4		2	
1				6			5	
								3

INTERMEDIATE - 128

	4							
			8				2	3
	6		2	5			4	
			1	3				4
		9				3		5
			5	4				6
	5		3	6			7	
			7				3	1
	2							

INTERMEDIATE - 129

		8	2		1			
					7	1	4	
1	5							7
6					4			8
		5		3		6		
3			8					1
5							1	6
	4	9	3					
			5		9	4		

INTERMEDIATE - 130

		1		6		4		
			7		3			
6	9		1		4		3	7
		7				5		
	4	3	5		8	7	2	
2		5				6		9
8								4
			2		6			

INTERMEDIATE - 131

				1			6	9
2	9							
		8				7		
		2			3		7	1
	6	4	1		5	8	2	
1	8		6			5		
		6				2		
							5	8
8	7			3				

INTERMEDIATE - 132

	6							9
		1		9			5	
			5				3	1
7	5			6	9			
8			2	1	5			3
			8	4			6	5
6	9				4			
	7			5		3		
2							8	

INTERMEDIATE - 133

3	1		5		8			2
		7		4				3
				3			1	
8								7
	3	6				4	5	
1								6
	7			2				
9				5		8		
4			9		7		2	1

INTERMEDIATE - 134

		4		8		7		5
					3		8	
6	1			7			3	
4		3	9					8
9					5	6		1
	7			5			2	3
	4		3					
3		5		1		4		

INTERMEDIATE - 135

				1				7
	2					9	5	
7	8		5	3		2		
	7		2			5		
			3		5			
		5			4		2	
		8		6	3		1	4
	4	9					8	
1				2				

INTERMEDIATE - 136

		1	5	2				
	9	2		7		5		
4		5					6	
						8	9	
	8			9			5	2
						1	7	
9		7					3	
	5	8		6		9		
		3	7	1				

INTERMEDIATE - 137

5			6		2			1
6				8				5
	1			7			4	
			9		5			
1	9		3		8		6	4
		3				5		
		1		5		2		
2		8	7		4	1		9

INTERMEDIATE - 138

	8		1					
	9			5			2	1
		2	3			9		
				1		6		3
	1		5		8		9	
9		8		6				
		9			1	2		
2	7			4			1	
					9		4	

INTERMEDIATE - 139

	3		7			8		
	6				8		5	4
5		2				9		
	1		2	9	4			5
			8		5			
8			1	7	3		2	
		4				1		8
2	9		4				3	
		8			9		4	

INTERMEDIATE - 140

8			1	7	3			9
	1						8	
7	4						5	1
1			9		4			5
				2				
9								6
6		1	5	3	7	9		4
5								7
	7			6			1	

INTERMEDIATE - 141

	7		6			1		9
		2		1				8
			4				2	
	8					7		2
		7				3		
3		4					6	
	4				8			
7				4		6		
8		6			2		1	

INTERMEDIATE - 142

			6					
7	5		8				1	
	8			7	3	5		
				4		3	5	
6								2
	4	1		3				
		7	9	1			8	
	2				8		9	4
					7			

INTERMEDIATE - 143

						8		
		1		5	6	2		
		7					6	5
	4	6	3				9	7
7	1	3						
	5	2	6				4	8
		8					5	1
		4		6	1	9		
						7		

INTERMEDIATE - 144

				9				
1			4		7			6
	7			6			3	
	9	7		8		3	1	
		5	3		6	7		
3	5	4	8		9	6	7	1
9	8						5	4

INTERMEDIATE - 145

	9			8		6		
	2							3
5		8			1			9
			8		7			4
	4			5		8		
			1		6			5
1		3			9			7
	5							6
	6			7		9		

INTERMEDIATE - 146

	9						3	
		7	5		8	9		
	8		4		3		6	
2								6
		9		8		7		
				1				
4	1						7	9
		5	9		6	8		
				2				

INTERMEDIATE - 147

	3	5	8					
4	1			7	9		8	
7			4				3	
					7			4
9								1
1			6					
	7				4			9
	4		7	1			6	3
					3	7	4	

INTERMEDIATE - 148

		1	2		3	5		
5			8		6			4
6				4				9
4								2
9	5	3				6	8	1
		9		3		4		
	1		6	8	4		2	
				5				

INTERMEDIATE - 149

			6					
7					4		6	
3				2			5	
5					7	4		
		8	4	3	1	5		
		3	5					6
	2			7				1
	8		2					4
					8			

INTERMEDIATE - 150

			8			2	6	
	4			6			3	
				4		5		7
		3			1			
	2		7			6	4	1
		8			4			
				5		4		3
	3			7			1	
			9			7	5	

INTERMEDIATE - 151

	8			2	6			
1		9						
	2		4			5	6	
		1						7
	5	7		1		8	3	
9						2		
	9	8			7		4	
						7		5
			3	4			9	

INTERMEDIATE - 152

			2	6		5		
			5		9	1	3	
					3			9
2		4		7		8		
9		1			6		4	
7		3		9		6		
					1			6
			6		2	9	1	
			9	8		3		

INTERMEDIATE - 153

		3		1			9	
	6	1	4			5		
7	4			3				
6		9				3		1
3		8				4		6
				5			6	2
		6			8	9	4	
	2			6		7		

INTERMEDIATE - 154

2		9		8			3	
				2				4
8		4	5					
			2					5
	4	8	7		5	3	2	
9					4			
					2	1		7
7				6				
	1			4		6		3

INTERMEDIATE - 155

			1	7	5			
				4				
8								2
	9						6	
	2	1				3	5	
	5		4	1	3		9	
3	4			5			7	9
	8	9	7		1	6	4	
				9				

INTERMEDIATE - 156

9						5		8
		2		9	5			
7					1		4	
	1	3		6				
	6		8		2		3	
				4		6	9	
	9		1					4
			9	3		8		
2		8						9

INTERMEDIATE - 157

		1	5		4	2		
5	9		7		1		3	4
			9		2			
	1			7			4	
7								5
3		6				7		8
	6						5	
	3	4				8	7	
			8	4	9			

INTERMEDIATE - 158

		6		3	9		5	
1			8					
		5		7		3		1
3							6	
2		8				4		3
	7							5
5		9		4		6		
					7			4
	1		3	6		2		

INTERMEDIATE - 159

		1	5	9			6	
				4				8
5						2		1
	3					8		
7		8	1		3	9		2
	4					1		
9						5		4
				5				3
		5	8	3			1	

INTERMEDIATE - 160

		6				8		
5			2		1			9
			3					
6	9		7	1			8	
	7		6		5		4	
	3			9	8		7	6
					6			
2			1		9			4
		7				6		

INTERMEDIATE - 161

					5			
	9			8			2	3
5	3		9					4
		9			8		5	
7	4				6			
		8			2		3	
2	8		5					1
	5			6			9	2
					3			

INTERMEDIATE - 162

	5	9	6					
				9	2			
1	6		5	8	7		9	
			4	3			8	
	9	4				7	3	
	3			5	6			
	7		9	1	5		2	8
			2	6				
					4	3	1	

INTERMEDIATE - 163

	2						3	
			8		2			
				6				
		6		1		9		
		3	9		6	7		
	4						8	
1			7	3	9			2
2								7
5				4				9

INTERMEDIATE - 164

	3	5						6
			6				9	
	1			9	5			7
		2			4		5	
		7	5	6				9
		3			9		1	
	7			5	2			1
			3				7	
	2	1						4

INTERMEDIATE - 165

7	2						9	3
	3			9			2	
8			4		7			2
	1						4	
4			8		6			5
	7	1	2		5	4	6	
	6		7		4		5	

INTERMEDIATE - 166

				4				
			1	2	8			
	2	7				8	4	
3	4			9			5	2
7								6
		5				9		
		6	9		2	7		
		3				5		
5				1				9

INTERMEDIATE - 167

				7				
2	7		6	3			9	
1			5		2	3		
7						9		3
		6				8		
8						5		2
5			1		7	6		
9	1		8	6			4	
				9				

INTERMEDIATE - 168

							3	
2	7	9			6	5		
				7	2			
4			7					1
		5		9		3		
7					8			5
			6	4				
		8	1			7	9	4
	1							

INTERMEDIATE - 169

	1	4	6		7	8	5	
			9	8	5			
				2				
4								8
8				5				7
	7	5	8	1	2	9	6	
2		7		9		6		1
	8	9		7		2	4	

INTERMEDIATE - 170

						9	4	2
	6				8			3
	9	4	1					
		7			4		8	
		2		7		6		
		9			3		2	
	8	1	9					
	5				2			1
						8	9	6

INTERMEDIATE - 171

		6			3		8	
		4	7					3
	3				6	4		
3			4		1			
1	8						5	4
			8		5			6
		7	3				2	
6					2	1		
	2		6			5		

INTERMEDIATE - 172

1	3						4	7
		9				8		
	4		9		6		2	
9		5				3		8
	2			3			9	
7		8		2		4		9
			8		9			
			4		1			

INTERMEDIATE - 173

		1		8			2	7
4	7		9			1		
			3			9		
2					3			5
						3		4
1					8			9
			4			5		
7	5		2			8		
		3		5			9	2

INTERMEDIATE - 174

						9		2
6	9				4	3	1	
1		8		2				
			6	1				7
				7	3	1		
			4	8				6
2		9		4				
8	1				6	7	4	
						8		3

INTERMEDIATE - 175

9	5	1	8				6	7
7								1
	3			2	7	1		
			6	9			4	
	6			3	8	2		
8								2
3	2	6	7				9	8

INTERMEDIATE - 176

8			3	4	2			9
		3	6	7	5	2		
9								6
	2			6			9	
	5						3	
5			4		1			7
		1	8		7	6		
3				2				1

INTERMEDIATE - 177

5			4			6		
			8	2			9	3
					6			
	6					5		
7			1		3	4		8
	4					1		
					1			
			3	5			1	7
4			7			3		

INTERMEDIATE - 178

		2	8				7	4
	8	1	4			2		
				2				6
	1	7		4				
			5		2			
				6		8	2	
1				8				
		6			9	3	4	
7	2				4	9		

INTERMEDIATE - 179

	1						7	4
	4			1				6
	2	6	4		7	9	1	
		7					4	
2				3				7
	8					2		
	5	2	1		3	4	9	
8				2			5	
1	9						6	

INTERMEDIATE - 180

				1		4		5
								8
7	9				5	6		
2					6		7	
		6	1	8	2			
5					4		8	
4	5				9	1		
								2
				4		8		9

INTERMEDIATE - 181

3		4			6			
1					3	2		
				8				6
	8			7				9
		5	1		9	7		
7				4			6	
5				9				
		7	5					2
			7			8		4

INTERMEDIATE - 182

9								8
			2		7			
	4			1			9	
2								1
8		9	1		6	5		2
5	2			3			1	7
		6		4		8		
3		8				9		4

INTERMEDIATE - 183

				5		3		
	8		2		7		1	
		4						5
5	1					7		
2					9			6
3	6					4		
		2						1
	9		4		5		8	
				1		9		

INTERMEDIATE - 184

						8	7	9
	6	1						
4				7		6	5	
				2			9	4
		3	4		9			
				6			1	7
9				8		7	4	
	5	4						
						9	3	5

INTERMEDIATE - 185

			3	8		6	7	
7	1							
		5			9			
	6		7		4		9	
		7					3	
	2		1		8		4	
		4			3			
1	9							
			4	9		8	6	

INTERMEDIATE - 186

7		9			3	4		
5				2			7	
			7		6	8		
4							1	
		5				9		
	1							6
		1	2		5			
	8			6				4
		2	3			7		1

INTERMEDIATE - 187

	1	8				2	7	
			8		1			
			2	3	9			
		3		8		9		
	8			6			2	
7		1				4		3
	9		5		6		4	
	5						1	
		6				7		

INTERMEDIATE - 188

4		3	9		6	8		2
		7				9		
1		9				3		5
8		5				4		6
	3			2			5	
			8		3			
		8		9		5		
	2			7			3	

INTERMEDIATE - 189

	2	6		7				4
			6					
	7			2	3		8	
1			7			8		
	3		8		2		6	
		2			4			5
	6		5	3			7	
					6			
2				8		1	5	

INTERMEDIATE - 190

4	2			7				8
		3			9			5
					4			
	1					8		4
8		4				3		9
3		9					1	
			8					
5			6			2		
1				5			3	7

INTERMEDIATE - 191

				6				
		3		9	1		8	
9		8					4	6
			5	7	9			
2	9	7			6			
			1	2	3			
7		1					9	8
		6		3	7		1	
				1				

INTERMEDIATE - 192

		3		6		8		
6	2				1			
		7					2	
	6			5				3
5					3	2		
	4			2				1
		9					4	
2	1				6			
		6		4		5		

INTERMEDIATE - 193

	7	1				9		
			7					
6				4	9			
7	5		9		2	3	1	
				7		8		6
4	6		1		8	5	9	
3				9	1			
			6					
	8	4				6		

INTERMEDIATE - 194

	6			1				
			8				6	
5	8		3					7
	2				1	8		
				2	9	7	4	1
	7				8	9		
2	5		9					4
			6				2	
	9			7				

INTERMEDIATE - 195

	4						9	
		1				8		
9	2						5	4
	1		3	5	4		2	
4	3			7			8	1
				1				
	8		5		2		6	
2			8		9			3

INTERMEDIATE - 196

9							1	
		6	8	5				3
	8		3					
7					2			8
		3	5		6			
1					9			6
	7		9					
		8	4	6				1
5							3	

INTERMEDIATE - 197

8				7				9
	1			5			4	
4		2		9		7		5
7		8				2		3
		9				8		
			8		6			
			9		7			
2								1
		5		4		3		

INTERMEDIATE - 198

9	7			5	8		4	
4		5		1		2		
	3		1		7	5		
	9			3	2	7		8
	2		9		5	3		
7		8		9		6		
3	4			2	6		5	

INTERMEDIATE - 199

					1	3		
4			7			1	8	
			8		9		6	4
8	1	6					7	
	4					6	3	9
1	5		9		4			
	7	4			8			3
		9	5					

INTERMEDIATE - 200

	1			5		9		
		4			7			3
7				1	2		6	
	6	7						
4		3		9		2		7
						6	5	
	3		5	4				1
6			8			5		
		5		7			2	

INTERMEDIATE - 201

	4	9		6		1		
		8	1				4	
				3	8			
9	2		5					
		1	2	8	7	9		
					1		2	4
			8	7				
	7				3	2		
		6		2		4	3	

INTERMEDIATE - 202

		5				8		
8							4	6
		6		4	9		3	5
				3		6	9	7
		4			6			
				5		4	1	8
		2		9	8		5	4
7							6	9
		9				3		

INTERMEDIATE - 203

		9	6				4	
1	2		8		9			7
	6			7				
		7	2			1		5
6		5			1	7		
				6			5	
2			1		7		3	9
	8				5	4		

INTERMEDIATE - 204

			8			5	1	
2		4						
		1			3	9		
4			1		2			3
	2		6		8		7	
3			9		7			2
		3	2			1		
						7		9
	4	6			1			

INTERMEDIATE - 205

	9			8	2	7		
		2	3	7				6
8			6					
		9		3	7		2	
			1					
		3		9	4		6	
6			7					
		7	9	2				4
	2			1	6	3		

INTERMEDIATE - 206

3			5	2	6			1
	5		7		1		2	
4								8
		2	8		5	4		
	7			4			6	
			6		3			
1		3				8		9
	4	9				3	7	

INTERMEDIATE - 207

		4		1	6			3
			7			9		6
			9				4	
9	8					5		
3	1		4		9		2	8
		5					9	1
	3				4			
2		8			7			
7			1	2		3		

INTERMEDIATE - 208

	7	5	4					
3		4	6					7
	9			1		3		
			2	4		8		9
							1	
			9	7		6		2
	6			2		4		
9		1	5					8
	5	2	7					

INTERMEDIATE - 209

		1				3		
9								8
	3	5				9	2	
			7		4			
	5		3		1		6	
	2						1	
			6		8			
1			9		2			5
	8	4				6	7	

INTERMEDIATE - 210

			1	3		8	4	
	2	1		5		6	7	
				7				
7							2	
	1	8	9		7			6
3							9	
				1				
	7	5		9		4	8	
			2	4		9	5	

INTERMEDIATE - 211

					4	1	5	3
7								
		1	6	8		2		
		7				6	1	
		2	8					
		5				4	3	
		8	7	1		5		
6								
					6	8	2	4

INTERMEDIATE - 212

4			1				9	
		8	9					
		9			3		2	
3					6			9
8			3				4	2
2					8			1
		6			7		5	
		1	4					
5			6				3	

INTERMEDIATE - 213

	6	9				7		
		8	3					
			5		6		8	
6	8			3			5	
			2		8			
	2			4			7	9
	7		9		1			
					3	6		
		4				9	1	

INTERMEDIATE - 214

		3					1	
						2	9	5
	9		4	7				
1			2		6		4	
	5		1		9			8
				1	8		3	
8	3	1						
	2					4		

INTERMEDIATE - 215

				3				
	9	3					5	7
			8				1	
4		8			9			
			7		5	6		
2		7			6			
			2				8	
	3	4					9	5
				9				

INTERMEDIATE - 216

4								9
	9	7				2	5	
1		5				3		6
			5		6			
			4		9			
5				3				4
		2				8		
	1		9	5	3		2	
	5		7		2		6	

INTERMEDIATE - 217

			1		2			
5		4		7		6		3
6				3				7
1								8
	6	7		8		9	3	
	8		6		3		7	
7	9						4	6
2		6				3		5
				2				

INTERMEDIATE - 218

	7			2		1		
			9		6			7
8		1	7			9		
	3					8	2	
1								3
	6	4					9	
		3			9	6		8
7			4		1			
		6		5			7	

INTERMEDIATE - 219

		2	4					
					1	6		
	4	1		3	7	8		2
	6	3						4
		8		5		7		
9						5	1	
7		6	3	4		1	2	
		5	7					
					2	3		

INTERMEDIATE - 220

		2		1		6		
				8				2
8	1		5			9	7	
5		8	7					
7					3	5		
2		3	1					
1	8		2			3	6	
				6				5
		7		9		2		

INTERMEDIATE - 221

	8	3				5		
	5		6				3	9
7		6				8		1
				1			9	
			3		4			
	6			7				
9		4				1		6
1	3				9		2	
		8				9	7	

INTERMEDIATE - 222

5			2	9				
	6		7				3	
						2	6	5
					5			8
	3	6	8		2	5	4	
7			3					
6	5	9						
	1				7		5	
				3	8			6

INTERMEDIATE - 223

						4		
7	9		2					
	3		6	7	1		5	
3				5				
5	4		7	8		9		
2				9				
	7		8	6	9		2	
9	6		3					
						8		

INTERMEDIATE - 224

		5				4		
			9	4	1			
1		9				3		6
	6						5	
8			6		2			4
	5			1			8	
		6	5		4	1		
			3		6			
4								5

INTERMEDIATE - 225

	3		2			8		
	2					6	4	9
8	5				6			
		8		2				4
			1		4			
4				5		1		
			3				6	8
9	7	3					5	
		6			7		2	

INTERMEDIATE - 226

		5			8		9	
1				2		3		
	3				4			7
2		1		8				
	7		2		9		5	
				3		7		8
3			4				1	
		4		5				2
	2		1			4		

INTERMEDIATE - 227

	2	8					9	
		7	1					
				9	3			1
2		3			4			
7								2
			7			3		9
4			5	8				
					1	2		
	1					7	3	

INTERMEDIATE - 228

6		8			2			
		5			4	8		
	2					4	1	5
					8		7	
		2	4		7	1		
	7		5					
7	8	9					4	
		1	2			5		
			8			9		7

INTERMEDIATE - 229

		1	3	5	6	8		
8	4						7	6
		9	5		2	7		
4								9
	5			6			4	
	8						1	
			9		4			
	1	6		8		4	9	

INTERMEDIATE - 230

	4					1	5	
7			5					3
5		3		2		4		
				6			3	
		4	8		1	7		
	9			5				
		8		7		9		2
9					3			8
	5	1					4	

INTERMEDIATE - 231

			3		5			
	6						3	
9								8
	3	8				6	1	
1				6				5
3		6		7		2		9
	1		6		2		5	
8				9				4

INTERMEDIATE - 232

		7	5				1	9
		8			1			
3	1			6				
	2			5	9	7		6
8		9	7	3			2	
				1			9	7
			3			8		
5	7				8	6		

INTERMEDIATE - 233

						3	7	6
		8	3					
			6		9	4		
		1	7		3	9	4	
4								5
	9	3	2		5	8		
		4	9		8			
					7	1		
1	3	5						

INTERMEDIATE - 234

			6	7	3			
		1		5				
6		2					9	
	2		9		5	1	3	
						4	8	
	8		1		4	7	2	
8		3					7	
		4		9				
			7	3	8			

INTERMEDIATE - 235

					6	7	5	
	4	8						
			8			1	9	
4			6	7	9		8	
	7	5					6	
8			4	1	5		2	
			7			2	1	
	2	6						
					3	8	7	

INTERMEDIATE - 236

			4	8	9			
		1				4		
4								5
		4	2		6	8		
	3		1		5		2	
	7						1	
		3	8		1	6		
2	5						4	1
			5	2	4			

INTERMEDIATE - 237

		7				5		
	1		8	6	7		3	
	3			2			7	
5	4		6	1	3		2	9
7			9		5			3
8								1
		2	1	8	9	4		
1								7

INTERMEDIATE - 238

5						7		1
	9	4	7					
				8	5	9		
		3		4				
4	2	1	6				5	
		7		2				
				3	7	4		
	7	6	1					
3						5		7

INTERMEDIATE - 239

	3	7		1		8	5	
8								9
	5						1	
2			9	5	4			6
5				2				1
		4		7		9		
	4						7	
			6		1			
				3				

INTERMEDIATE - 240

6	3	2		7		8		
			3	5				
		8			1			
8					6		2	7
4	1		5					9
			2			4		
				1	3			
		1		4		6	7	2

INTERMEDIATE - 241

4	7				2			
		9	5					3
		6					7	
					1			7
7		8	3				5	
					9			1
		4					3	
		1	8					5
9	5				6			

INTERMEDIATE - 242

	6		5				7	
7	3	1		8				5
	1		7				6	2
		5	2		6	9		
9	2				1		3	
1				9		8	4	7
	9				5		1	

INTERMEDIATE - 243

8		6	3		5		7	
		7		2				
	3	1						5
				7	4		5	
6								
				8	1		6	
	6	9						1
		8		9				
7		5	2		6		8	

INTERMEDIATE - 244

2			3	8				1
	8		1			6		
5					7			
					6	3		
		4				7		
		5	9					
			7					9
		1			4		7	
6				5	1			2

INTERMEDIATE - 245

	6	5		4		3	7	
	3		9		6		2	
3	1		5		9		4	2
	2			3			8	
		4				9		
		2				1		
		1	2		4	6		
5								7

INTERMEDIATE - 246

	2		7					
3				5			7	
	8	4	2				9	
		8		7	3	5		
					8			9
		6		2	9	8		
	5	7	1				3	
1				8			6	
	6		3					

INTERMEDIATE - 247

	6			5			4	
			2	9	4			
2		7		8		5		3
	5			2			8	
9								4
4								6
	3		5	4	2		1	
6	9		8		7		2	5

INTERMEDIATE - 248

		1		7		8	3	
					6			
	8							9
		7				6	8	
8	2			5			1	7
	9	3				5		
4							2	
			2					
	5	9		8		4		

INTERMEDIATE - 249

				6	2		8	
	6	9					5	
						7		4
7				2				
	5		9	8	3		7	
				7				1
4		5						
	7					9	3	
	2		8	9				

INTERMEDIATE - 250

1								5
			9	1	5	2		
	8		4					
	5			4		3	8	
	6		1		9		2	
	2	1		7			6	
					6		7	
		2	7	9	1			
8								3

INTERMEDIATE - 251

				1	9		6	7
		1				5		
2				3			1	
	7	3						1
1	6						4	9
4						3	7	
	1			6				5
		6				4		
5	8		3	2				

INTERMEDIATE - 252

	1	9		3		5	4	
5		7		4		6		3
			3		4			
8	5	3				2	9	4
	9						5	
	3			2			8	
4		5	6		8	3		2

INTERMEDIATE - 253

	9	7	1		2	3	6	
			3	9	7			
		3				1		
		8		3		9		
9		6				4		2
1								3
7								6
4			6	1	8			5

INTERMEDIATE - 254

		9	4		3	5		
	1						8	
4		6	5		9	1		2
8			2		5			1
		2				6		
5				7				8
	5						4	
9		1				8		5
	4		8		2		1	

INTERMEDIATE - 255

9			7			3	8	
8								
		1	8	9		6		2
				7	9	5		
			2	3	4			
		9	5	1				
1		5		8	7	9		
								4
	9	7			3			5

INTERMEDIATE - 256

	7			1		8		
		8	6					3
6					7		4	
		6	5		1		8	
8								9
	2		7		9	5		
	8		9					4
3					5	7		
		5		4			9	

INTERMEDIATE - 257

5	2							7
		7						
1		9	4				6	5
				3	1		8	
3					9		7	
				4	5		1	
9		1	2				4	6
		8						
2	3							1

INTERMEDIATE - 258

	3			5	1	2	7	
				6		9	4	
		5						
	1			8				6
	7			9			2	
5				3			1	
						3		
	4	8		7				
	6	1	2	4			8	

INTERMEDIATE - 259

5			9			8	4	
	7							
			1	8				2
1					3	9	7	
	6		7		1			
9					8	1	5	
			2	5				8
	1							
2			8			4	9	

INTERMEDIATE - 260

9				3	5			7
					4	9		
	2		1	9				
1	6					5		
8		3				4		1
		4					3	9
				5	6		9	
		8	4					
2			3	1				4

INTERMEDIATE - 261

		9	4		7			
				8	1		6	5
				9		1		
8							5	2
1						4		
5							8	1
				4		7		
				7	2		4	6
		4	8		9			

INTERMEDIATE - 262

7				9				
1	4		2		7			
2	8		3		5		1	
		1				6		
	6			5			9	
		2				8		
	2		4		3		8	9
			9		8		7	5
				7				4

INTERMEDIATE - 263

		1	4					9
4		9						3
					6	4	5	
8			2	3				
					7		9	
6			1	9				
					1	2	7	
7		8						5
		5	6					8

INTERMEDIATE - 264

2	3			9			5	6
8			1		4			2
		4		5		1		
5			6	8	7			4
		9		2		8		
			5		6			
	8						6	
9				4				1

INTERMEDIATE - 265

	9		8		2		6	
		3	6		5	7		
1								3
	6		4		1		7	
4	2		9		3		1	6
		1				4		
	7			5			8	
6				9				2

INTERMEDIATE - 266

					2			8
		3					2	
	2			5	7	6		
3	8							5
	5			8	1		7	
6	4							2
	7			6	8	5		
		6					1	
					4			3

INTERMEDIATE - 267

			1	8				
		6			9			
		5			6	2	1	
	1	4		3				8
2			7		8			1
6				2		5	4	
	9	1	5			7		
			3			4		
				9	7			

INTERMEDIATE - 268

6				4				1
		4	1	7	9	5		
	8						9	
		5				8		
			5		8			
8			4		1			3
			9	6	5			
5	2						8	9
	9						4	

INTERMEDIATE - 269

4								5
		2	3	6	4	9		
1		6		2		4		8
		1	8		2	7		
3				7				1
6	2						5	4
7								2
			4	1	3			

INTERMEDIATE - 270

	7				3	8		5
		8	2	4	7			
				5		7		3
2			5	7	6			
8			3	9	4			
				2		6		7
		1	7	3	9			
	4				8	3		9

INTERMEDIATE - 271

	3		2	1	8		7	
		5	7		4	6		
	4						2	
			8	5	6			
	1						4	
6			3		1			7
3								2
		1	5		2	9		

INTERMEDIATE - 272

3		9			2	6	5	1
2		6						
1							8	3
5			6	4	9			
			2		5			
			1	3	7			5
6	3							2
						8		6
9	2	8	5			4		7

INTERMEDIATE - 273

1			2				9	4
		7				5		
	2				5			7
	9			6				
				4	9	8	5	
	3			2				
	4				1			8
		1				2		
9			7				1	3

INTERMEDIATE - 274

8	4					9		6
		6						4
9		2			4	7	1	
		7	6		8			
				5				
			9		2	6		
	1	9	2			8		5
6						1		
3		8					9	7

INTERMEDIATE - 275

		5	8				3	
		8		9			4	
		1		3	2			6
			2	8	7	3		
7								8
			5	4	6	9		
		7		5	4			3
		2		7			9	
		4	6				8	

INTERMEDIATE - 276

1					3		8	4
6		2						
	8	3			6			9
				9			6	8
7	2		6					
				8			7	2
	6	9			4			5
8		1						
5					8		9	6

INTERMEDIATE - 277

				6				
2		1				5		6
		4	5		3	9		
8		7				2		4
				5				
	6	5		9		7	8	
1				8				9
7			9	2	4			5

INTERMEDIATE - 278

	2	9	6		8	1	5	
8		7		1		2		4
		6		2		9		
		5				8		
4			2		9			6
		2		8		7		
	9		3		4		1	

INTERMEDIATE - 279

		9			1			
	4					2	3	
		6	5	4				9
			1		8			6
6		3		5			4	
			3		4			1
		8	9	7				2
	9					7	1	
		2			5			

INTERMEDIATE - 280

6	2		5		3		4	1
3								5
			1		9			
		8				6		
	3						7	
		7		9		5		
8	7			1			5	9
1				4				2
				2				

INTERMEDIATE - 281

6			2			4		
	9				4			3
4	5			9				
			5		9		2	
					7			5
			6		8		3	
8	7			1				
	6				2			7
1			8			5		

INTERMEDIATE - 282

			1					7
6		8						
		1	7	5	8	6		
	9		3	7		2		
3			4					
	2		6	8		1		
		7	5	6	9	4		
2		3						
			2					5

INTERMEDIATE - 283

		4				8		
1	3			5			2	7
6				3				4
2			5		8			9
	8						4	
			3	1	7			
	5		9		3		6	
	1	3				2	7	

INTERMEDIATE - 284

	8		5	4				9
						1		
		3					7	5
6	9				3	4		2
				9				
8		7	4				9	3
5	6					2		
		4						
7				6	4		5	

INTERMEDIATE - 285

				8				2
	2	1		5	4	9		
5			9					4
	4	5						
			8		5			
	1	7						
3			1					6
	9	4		6	3	5		
				2				1

INTERMEDIATE - 286

	3						4	
8	4						6	5
			8		4			
	7			1			3	
2								1
		9				7		
		1		9		2		
6			7		2			3
	9		6		1		8	

INTERMEDIATE - 287

6							8	
4	9							6
					9	3		
9	6		5	2			4	7
	7		4					
5	8		7	6			3	2
					2	5		
7	2							3
3							9	

INTERMEDIATE - 288

	4		6			7	3	
		2	4					9
	1		3				4	
		3						4
	5		7		4		8	
7						3		
	3				1		9	
4					3	5		
	2	8			5		1	

INTERMEDIATE - 289

		8		6		5		
	7		1		9		8	
			9	5	1			
	8						2	
		3				1		
	2		8		6		5	
3		4	5		2	8		1
8	5						3	2

INTERMEDIATE - 290

		6				9		
1		2	5		7	6		4
			3		1			
		8				2		
6	2	7		8		3	5	9
2								6
	4		9		6		7	
		3		4		1		

INTERMEDIATE - 291

	6			9			3	
	1		8		3		6	
	8	3	7		4	9	1	
6	9			2			8	3
				1				
		5		7		1		
				8				
1	7						5	9
	3						2	

INTERMEDIATE - 292

8	2		9	7		3		
							6	
			6		3	8		5
		3		4			2	
9								3
	8			5		6		
7		9	4		8			
	5							
		4		9	6		7	8

INTERMEDIATE - 293

				9				
	2			3			4	
8								6
	3		1		5		6	
		4	6		3	5		
1								2
			8		7			
7		5				2		8
	6			5			3	

INTERMEDIATE - 294

							9	
	3		7	1				6
					8	2		7
9	1		8				3	
	4							5
7	8		4				6	
					6	9		8
	7		9	5				4
							2	

INTERMEDIATE - 295

				7				
2			4		3			8
	4		1		6		7	
			2		9			
		9				3		
	2	5				1	9	
	1		8	2	5		3	
	8		7		1		4	
				9				

INTERMEDIATE - 296

2		6	5		7	4		9
				2				
		8				5		
	5	7	1		4	6	2	
6								8
		4				1		
			2	8	9			
	8		4	1	3		5	

INTERMEDIATE - 297

5				7			3	
	1	6						4
7			4				6	
		8	5			6		
			7				4	
		1	9			8		
6			8				9	
	9	4						2
1				2			5	

INTERMEDIATE - 298

								4
	3		5			6		
4		5	8				3	
					3	2	4	
1			2	8	6			7
	9	8	7					
	5				1	4		6
		6			5		2	
7								

INTERMEDIATE - 299

			3			6	1	
	8						3	7
		3	4					
		5			4			6
4	3		6	9				
		2			3			9
		8	9					
	5						9	2
			1			7	4	

INTERMEDIATE - 300

	7					2		
1				5				
		9		1		3	7	
	2	5			7			3
								9
	3	7			9			6
		3		6		5	4	
7				3				
	4					6		

SUDOKU
PUZZLES
EXPERT - {1 -500}

EXPERT - 01

	4		8					
2			3					1
		5			2		7	
		4		9		5		
3	8							7
		1		2		8		
		7			6		3	
9			5					4
	2		9					

EXPERT - 02

3	1						4	6
					1			8
		4			3	1		2
		5		8	2			
			7		6			
			5	1		4		
7		1	9			8		
9			6					
5	8						2	9

EXPERT - 03

	4		5		6		8	
3		1				5		9
		2	4		8	6		
		8				2		
	5		2		1		9	
		5	3	7	4	9		
8			6	9	5			3

EXPERT - 04

			3	8	7		6	
7						9		
	2	6				3		
1			9	3	8			4
4			6		2			9
9			4	1	5			6
		5				6	7	
		1						3
	7		1	4	3			

EXPERT - 05

5			4		7			9
			5		1			
9	1			4			7	2
		4				9		
		2	6		9	8		
	8	3		2		6	4	
	2						3	
4								8

EXPERT - 06

	4			2			7	5
2				4		8		
	1		9		8			
		5					3	1
	6							
		2					4	7
	5		2		4			
3				9		1		
	7			6			9	2

EXPERT - 07

2	9		6		4		1	8
	7	6				4	9	
			1		7			
		2		6		8		
	4			5			7	
6			4		5			3
	1						2	
4			2		8			7

EXPERT - 08

			8		1			
7	2	4		9		1	3	8
9			2		4			6
1								2
	8	3				7	4	
	6						8	
		5		4		2		
				2				
6			5		7			3

EXPERT - 09

8	4		6		7			
				5		6	9	1
		6	7					3
		2	5	6		1		
		9	3					2
				3		2	4	7
3	6		9		2			

EXPERT - 10

							2	5
	3	8	9	1				
		5				3		
		2	1		8			3
8	7				3		4	
		3	2		6			7
		1				4		
	2	7	4	3				
							3	1

EXPERT - 11

			3		1			
1	4			7			5	
						1		
6		9	8			5		1
				2	6			7
7		8	1			9		6
						4		
9	1			3			8	
			7		5			

EXPERT - 12

			7		5			
7								3
	1		9		2		7	
3	9		8		1		4	7
		1				8		
	5						3	
		8	2		6	9		
	2	7		9		6	1	

EXPERT - 13

9			5		6			3
		8	1		2	9		
		4	8		9	5		
			4		1			
4		6		5		1		7
8								2
		7				8		
	6						7	
	8	9				4	3	

EXPERT - 14

	9	6			4			
						7		
				5	7		6	1
6				7	2	5	4	
	7	5	1	6				3
1	4		5	2				
		8						
			3			2	1	

EXPERT - 15

				1				
9		8	5		7	2		3
	7		9		4		5	
4								6
				2				
		9		7		5		
	8			4			7	
		3				9		
2			6		1			5

EXPERT - 16

4					9		2	
			8	3				5
								8
	1				2	5		4
		4	1		6	7		
2		9	7				1	
7								
3				9	4			
	6		2					3

EXPERT - 17

	8		3					
2	6		7			5		
		4			1	9	7	
				9		1		5
8			5					3
				8		7		6
		3			4	8	6	
9	4		8			3		
	2		9					

EXPERT - 18

4				2				
	3				6	5		7
6					3	2		
			6				3	
9	2			4			6	8
	6				2			
		2	9					6
5		4	1				2	
				7				4

EXPERT - 19

	5			3		7	6	
				7				
9			2		1	3		8
	2		1		3	8		
				2				
		3	8		4		7	
2		5	7		9			6
				8				
	3	7		1			8	

EXPERT - 20

			9		5			
8								4
	2						7	
	4		7		1		9	
	5			4			8	
	7			8			4	
		4		9		5		
	8		6		3		2	
3			4		7			6

EXPERT - 21

	5			3			6	
	7						9	
	9		4		2		5	
			2	4	8			
	3	1	7		6	9	8	
5	1						3	9
	2						7	
		7	1		5	2		

EXPERT - 22

1				5		2		8
		6			1			
5			3		8		6	
	7	3				8		
8				9				6
		1				5	9	
	9		6		3			5
			5			9		
2		5		8				4

EXPERT - 23

		7				9		
9								2
			5		4			
		4	9	8	6	3		
7		9				8		5
	7	2	3		8	5	4	
				1				
8			4	2	9			1

EXPERT - 24

5				8				2
	1		5			9	8	
		8		3				
					3	8	1	5
				1	2			6
					4	2	9	7
		1		7				
	4		1			7	6	
7				6				8

EXPERT - 25

	9	7			3			
6								
	8	3		1	2	5		
	3			7				
	5		4		8		3	
				9			6	
		1	6	8		9	4	
								7
			9			6	1	

EXPERT - 26

		5					9	
	2		5				4	
		8		1	9			7
					5	4		
1			7		6		2	9
					3	7		
		4		5	8			3
	3		9				8	
		6					5	

EXPERT - 27

	8	5				6	7	
			5	8	6			
5	3		8		7		6	4
		8	4		1	7		
	7						2	
2		1		7		5		6
	6	7		4		2	3	
	5						8	

EXPERT - 28

	3						9	
4	9				1		3	2
		1			2	8		
	8	3		2				
			3		9			
				5		7	6	
		6	2			4		
2	7		4				5	1
	4						8	

EXPERT - 29

6				5	2		4	
	1		3					
				9	4	7	3	
		9	8					4
8							9	
		7	4					3
				8	7	2	6	
	7		5					
9				3	1		7	

EXPERT - 30

	3						9	
9				2				4
5		7				8		1
		6				5		
3	9			8			4	6
8	7	5				9	1	3
7			4		2			8
	8			3			6	
			1		8			

EXPERT - 31

		1				3		
4								7
7		8		4		6		1
6			8		7			4
			1		2			
8		2				7		5
	5			6			2	
2			9		1			3
		9				1		

EXPERT - 32

	3		5		6	8		
8	5				3			1
		7		8				
7							5	
		3		1				
9							2	
		8		4				
6	7				9			3
	2		6		1	5		

EXPERT - 33

	7			4	6		1	
4			8					9
		2				4		
7	1			8		6		
		6						
8	4			3		5		
		7				8		
6			7					4
	5			2	8		3	

EXPERT - 34

6					8	2		
		9						3
				9	2			5
8	7	5	1	4			6	
	3			6	5	9	1	8
5			4	3				
2						4		
		3	2					9

EXPERT - 35

8			6				9	5
	3	6		5				1
		7						4
			5					
5	1	2				7	8	3
					2			
3						4		
7				8		5	1	
1	2				6			8

EXPERT - 36

5				6	4		1	
		8	2					4
	1					5		8
	8		5		1	2		
			6	4				
	3		7		2	9		
	4					7		5
		5	4					2
2				5	7		8	

EXPERT - 37

7		1		8		6		4
							1	9
					6	8		
	4		2	6		7		
			7		5			
	1		3	9		2		
					7	9		
							5	2
6		5		2		4		7

EXPERT - 38

1				9				
		8		7	3		9	6
2			1					8
		1	3	8			5	
	8					2		
		2	7	4			8	
7			4					5
		5		3	9		1	4
3				5				

EXPERT - 39

			9		4		3	
		7		3				6
	4					7		
		2		8				3
9			6					5
		6		5				9
	5					2		
		1		2				8
			1		7		5	

EXPERT - 40

					2	6		
	7	8				3		2
2			8					7
	5		7	4				
7				2				6
				8	5		4	
1					3			5
9		7				2	6	
		5	4					

EXPERT - 41

5					2			7
	1			7			2	
		8	9					1
6	3			5				
						6	3	
4	8			6				
		6	4					3
	9			8			5	
7					9			8

EXPERT - 42

	4		5				2	
8			3	1				
		7		2				3
1	2							7
			2	9				
7	5							9
		5		6				8
9			8	5				
	8		9				3	

EXPERT - 43

7			5				3	
		2			9	1		
				4			2	7
	7		8					2
		6				9		
2					1		4	
5	3			1				
		4	3			8		
	1				7			3

EXPERT - 44

8			3	5	6			
5		9					1	
	7	3	8					
				6			5	
		6	2		3	1		
	4			8				
					7	3	2	
	3					5		1
			6	3	1			4

EXPERT - 45

					2			
	5		3	9		2		1
	2			5			4	
		7	4			5	3	
8			9					
		1	5			7	2	
	1			4			8	
	7		6	8		4		2
					5			

EXPERT - 46

	1	3				5		
		4			1			9
	5	7	9					4
		9	1				6	
			4		8			
		8	7				1	
	3	1	6					8
		5			4			3
	9	6				7		

EXPERT - 47

8				6				9
	3		1		5		8	
	5						2	
		1				9		
7	4						3	6
			7	5	6			
4				9				7
				2				
	9	3				5	4	

EXPERT - 48

					1		9	4
		3		8		2		
	7		6					
		9			3			5
2					5		7	
		7			9			6
	3		9					
		4		1		9		
					6		5	7

EXPERT - 49

		3		6		2		
6		1		2		5		8
	5	2	7		1	3	6	
3	9						2	7
			8		2			
	2						5	
4		8				7		3
			9		7			

EXPERT - 50

				5	8			
7	4	8	3					2
	1				2			
	7					4	5	
6			1			9	8	
	9					3	2	
	6				4			
2	8	1	5					4
				2	1			

EXPERT - 51

3	4					9		
1	5			7	6			2
			8		3			1
					2	4		9
	2			6			3	
					9	6		8
			6		4			5
6	7			8	5			3
5	3					8		

EXPERT - 52

	5					7	8	4
3	8							1
				1			5	
		8	7	3	2			
				5				
			9	4	8	1		
	3			9				
9							6	2
5	2	7					1	

EXPERT - 53

2			1	6	3		4	8
	9	3	2					1
	8		3		4			
		4	9				8	
	2		7		6			
	7	1	4					6
4			6	2	7		1	9

EXPERT - 54

					1	7		
	1	2	7				9	
5		9				1	8	
8				4			2	
			5		2			
	2			8				5
	4	5				8		3
	8				3	6	5	
		7	8					

EXPERT - 55

	9						6	
		6	7		4	9		
	3			5			7	
6								7
			5	8	2			
9				7				8
2			4		5			3
		9				5		
4	1						8	6

EXPERT - 56

2			4	6	5			
4	5			3				8
		3					6	
	3				6	2		
				4			5	
	9				2	8		
		1					9	
3	8			5				1
7			9	1	4			

EXPERT - 57

	9		4		8		3	
1			6	7	3			4
9		3		6		7		8
			3		9			
	8			1			5	
8	1						4	5
4		7				8		9
		9				6		

EXPERT - 58

	2					1		
1			9				4	8
		4					7	3
				8	7	3		
		9		2	5			7
				9	4	8		
		3					5	2
7			4				3	6
	5					4		

EXPERT - 59

	6	7	4					
	5			6				
						6	4	7
	9				2	8	6	
			5		9		3	2
	3				6	9	5	
						5	1	6
	8			2				
	1	9	3					

EXPERT - 60

8		3			5		6	9
6				8	4			
			3					5
3	6					8		
	8			2			3	
		5					9	2
5					7			
			8	1				7
7	1		6			3		8

EXPERT - 61

		6	2					9
3		8			7			
4						3		
6			4					3
	3		7	2	5		4	
2					1			5
		5						1
			5			4		8
1					4	2		

EXPERT - 62

6			2		5			4
2								8
				7				
	3		6		4		8	
	2	7		3		5	9	
7	8						4	9
1			8		7			5
	6						3	

EXPERT - 63

			5		6		7	4
	6		1					3
						9		
	2	4		9			3	
	8			5			4	
	3			6		8	1	
		6						
3					7		2	
7	1		6		4			

EXPERT - 64

			8					
					6	4	8	5
7		8	1					9
	8			5				1
		1			4	2		
	7			3				6
4		6	5					8
					7	1	9	4
			4					

EXPERT - 65

	3		6		4			
6	4							1
					8		5	4
						3	4	
4		2	8		3	9		5
	5	3						
8	9		5					
1							9	7
			9		1		2	

EXPERT - 66

3		9					6	4
6			1					
				8	6			3
		6		3			5	
		5	6		7	2		
	7			5		9		
7			8	6				
					4			5
2	9					6		8

EXPERT - 67

	9			6			8	
8		2						5
			3				7	
			7		2	3		
5								9
		1	4		9			
	6				5			
4						1		6
	5			4			3	

EXPERT - 68

	4	2			3	7		
9			8		4	1		
3	1							
	2		4					5
6			9		7			4
	9		3					7
4	8							
2			6		8	4		
	6	9			5	2		

EXPERT - 69

9				8	3		4	
		5						
				4				8
4				6	8	7		
2		8					9	6
5				3	2	8		
				2				5
		4						
1				9	7		2	

EXPERT - 70

	7	1	3				5	
5		2		9	1			7
							2	1
	1		8		5			2
	5						1	
7			6		2		9	
3	4							
1			7	5		2		8
	2				9	5	6	

EXPERT - 71

		4	1		8	3		
		8		3		5		
9								2
	9						4	
		7				9		
5		6	3		9	1		8
3				9				1
	1	9	4		3	7	8	
				8				

EXPERT - 72

6	8							
		4				5		
		3	7	2				6
			8	6			1	5
	4		5		1		9	
5	1			4	9			
2				7	6	9		
		5				2		
							3	1

EXPERT - 73

3			4		5			1
5	6						8	7
				8				
	8		7		6		3	
				4				
6				3				8
	3		6		2		1	
			1		8			
	1	2				8	9	

EXPERT - 74

9							3	1
5	8		3				4	
		2			1	9		
				8	6		1	
	7	3						
				4	3		7	
		9			4	3		
8	3		2				6	
7							5	4

EXPERT - 75

	9				2		5	
	1							4
5		3			9			
	4	8					3	
		1	3		8	4		
	2					8	9	
			5			1		3
8							2	
	6		2				7	

EXPERT - 76

		4			5		7	2
		5		2				8
			9	8		6		
2		1					6	
	7						2	
	4					9		7
		7		3	9			
1				4		7		
4	9		1			8		

EXPERT - 77

			8	6				7
	7	8	1					
6	9			3				
	2			9	5		3	
		6						9
	5			1	4		7	
2	1			5				
	8	7	3					
			9	7				3

EXPERT - 78

		9				2		
8			1		4			5
			2		5			
	2	6				7	1	
		4		2		6		
				6				
			3	1	9			
	9		4		8		2	
	1	7				3	9	

EXPERT - 79

3								9
5			4		7			3
		8	1		3	4		
	8						6	
4			6		2			1
6								2
			9	2	5			
	5						7	
	3	6				9	1	

EXPERT - 80

		5	4	1		7		
7				5	8			
2	6							
		6	5				8	
8			6		3			9
	2				1	3		
							3	5
			9	3				1
		3		4	2	6		

EXPERT - 81

3			4		2			1
4	8			3			9	5
			9		6			
1								7
8			2		9			6
		9		7		2		
	6			9			7	
		4				3		
	3	8				1	6	

EXPERT - 82

	9	4				8	7	
1								9
2			9	7				1
				1		6		
		5	6		8	7		
		9		5				
4				8	2			7
9								4
	7	2				1	3	

EXPERT - 83

	2						5	
4			6		9			7
	7						3	
				4				
	3		2		5		9	
5		8				2		1
3	4			5			1	6
				9				
	6		4		1		7	

EXPERT - 84

			1	6		8		
7	1	8	9		3			
			4					1
8	6		7			9		
	3				2			4
1	7		3			2		
			5					9
5	4	1	8		9			
			2	3		4		

EXPERT - 85

	6			7				1
4		5	2				7	
1		3	9	8				
				2		5	8	
		6	3		8	7		
	8	4		5				
				1	9	2		7
	4				2	1		8
2				4			6	

EXPERT - 86

		6	1		9			
5						7	8	
	2				4			1
		7		4				
			2	8	6			
				9		1		
7			5				4	
	8	4						5
			4		7	3		

EXPERT - 87

	8			1	3	9		4
				6				5
5			4					
	7	6				2		
3	4					1		
	5	8				6		
4			9					
				2				7
	2			7	4	5		1

EXPERT - 88

		8				1		4
3					9		2	
	7		2	4				
7				2			8	
		4	3		6	9		
	8			5				3
				7	3		4	
	2		8					6
1		3				7		

EXPERT - 89

	6			7			1	
1		3				5		6
		8	3		5	1		
2				1				4
	5		2		7		6	
3				2				5
		6	1		9	3		
		4				2		

EXPERT - 90

6	3			5				
8		2				9		
	4		8					
				6		7	5	
		4		8	5	6	3	
				2		8	4	
	9		5					
1		8				2		
2	6			1				

EXPERT - 91

7				2		5		1
		1			3			9
2				1		4		
		6	1	3				
9			2		7			4
				6	9	8		
		8		9				7
1			8			9		
5		9		4				8

EXPERT - 92

				3	7	9		
							7	
5			8		4		3	2
1		2					8	3
				1				
3	6					2		5
8	2		4		1			9
	5							
		9	6	7				

EXPERT - 93

	9	4		5		3	2	
	8			3			6	
3		7				1		5
		8	5		9	2		
	2		8		6		7	
	5		1		7		3	
				8				
		9	2		3	4		

EXPERT - 94

1		8		3		2		9
	9		1		4		5	
			2		9			
5		1				4		3
	7						8	
				9				
		5		1		9		
		7				8		
9	6			4			3	7

EXPERT - 95

4	6		8		7			
		3		4	1			2
		1						
	1	9				7	2	
			7		8			
	3	7				9	4	
						2		
1			3	7		4		
			9		6		8	5

EXPERT - 96

	9							
6					3		5	4
	4			8		3	9	
1	2			9	7			
		4						
8	6			4	1			
	3			6		9	7	
5					4		2	1
	1							

EXPERT - 97

3			7		6	8		4
	9			4		6		
		8						
	1	9		7				
	5	6				1	4	
				6		9	8	
						7		
		2		1			6	
6		5	8		3			2

EXPERT - 98

	7		2	6	3		5	
		6	8		7	1		
	8		9	2	1		3	
2			6		4			5
3	4						9	6
7		4				9		1
			7		2			

EXPERT - 99

	2				9	6		
		8			4			
7					6	1	2	
		5		7			9	
			9					3
		6		4			8	
2					5	9	4	
		3			1			
	6				7	2		

EXPERT - 100

	5			6			3	
		9	7					6
		2	4		5			
	9	6						
7	3						6	1
						5	8	
			1		6	2		
4					7	6		
	6			8			9	

EXPERT - 101

	9		1					
8	7	4			6	1		
		3	4					
		7					3	1
9	1	2	6		3	4		7
		6					9	2
		8	9					
7	6	9			1	3		
	3		8					

EXPERT - 102

8	4				5		1	6
3			6					5
				9				
7			2		6		5	
		6				1		
	5		3		8			7
				3				
6					7			3
5	3		9				4	1

EXPERT - 103

			2					
	9	4		7		6		
5		7		4			3	
		5	6			4		
			1	8	2		7	
		3	5			1		
3		8		2			6	
	4	2		6		5		
			9					

EXPERT - 104

		3	7					
6				1	3		7	
8				4				
4		2			6			
	8	7		3		6	5	
			8			4		2
				8				1
	4		6	9				5
					4	9		

EXPERT - 105

	8		3	7	5		2	
6								9
	4			5			6	
7			8		6			5
	1		4		9		8	
3		2		8		7		6
4	7			3			5	8

EXPERT - 106

		5		9			2	8
			8			5	4	
		3		5				6
					8		9	
9				6				3
	4		7					
5				8		2		
	8	7			4			
1	9			7		8		

EXPERT - 107

1			3			7		
		4		1				8
5					4	2	1	
4					7			
	2	1		9			6	
3					1			
8					2	9	3	
		2		3				7
6			7			4		

EXPERT - 108

		9	3			2		
6	2						5	9
1					5			
			1			6		
	6		4	8	2		9	
		5			3			
			2					7
8	9						3	2
		2			1	5		

EXPERT - 109

7			4	9		8		
8				1				
	2			6			3	
						5		
1		4				9		2
		8						
	5			2			6	
				4				1
		2		8	6			9

EXPERT - 110

				9	8			
		6	5				2	
4				6		3		
	4		6	8				1
					1			
	8		9	3				7
9				4		1		
		1	3				6	
				7	6			

EXPERT - 111

	9			8	6			
3						9	1	
		2	3			4		
		9		4	2	1		
7								4
		5		6	9	3		
		7	6			8		
4						5	7	
	1			3	7			

EXPERT - 112

	7				9			
6			4	8			3	
		8	6	7			1	
9						4		2
		2						
4						3		1
		5	1	3			4	
1			5	9			2	
	6				7			

EXPERT - 113

			1		9		5	
		6		2		1	4	
		5			3			8
	6	9						3
		8				7		
7						5	6	
2			6			8		
	8	1		9		6		
	4		2		7			

EXPERT - 114

							3	
			8			6		1
8		9	5			4		
	9			3			1	6
5	1			2			7	
		8			4	1		2
4		7			1			
	6							

EXPERT - 115

		8		7	2			9
1				6		2	8	
							6	
7				3		9		
	6		9			5		
3				2		6		
							4	
2				1		3	9	
		4		8	3			2

EXPERT - 116

	1		4			9		
		4			8	7		3
9	7				1		4	
	5	2						1
				6				
7						4	5	
	4		8				3	9
6		8	3			2		
		7			2		8	

EXPERT - 117

		1	8		4			
9	6				1			2
	3	8						9
				5			9	
		7	4			6		
				8			2	
	1	4						3
6	5				2			4
		2	5		8			

EXPERT - 118

	3		4		5		8	
	8	9				4	1	
6			5		2			8
		4	7		3	2		
		6				1		
				2				
3	4		1		8		6	9

EXPERT - 119

		9	6		1			3
	8			5	9	6	4	
			4					9
	6					1		
			5	9	4			
		5					9	
8					5			
	7	2	9	6			8	
5			1		7	3		

EXPERT - 120

7		4			1			9
			3			2		
	9		4					8
1				3		8	2	
			8		9			
	8	3		2				5
9					8		7	
		5			3			
2			9			6		1

EXPERT - 121

	9			2	1	5		
			7					1
1			5		9			
4		5					9	
				3		6		
9		7					3	
5			2		3			
			6					8
	1			5	4	3		

EXPERT - 122

	6						9	
4		7			2		1	3
	3	1	5					
					4			7
7		9				6		1
					1			5
	2	8	1					
1		6			9		7	8
	7						5	

EXPERT - 123

		6	3					7
					7	9		
5			8					
9			5			8		4
6	5			3			2	9
8		4			2			3
					9			2
		8	6					
2					3	7		

EXPERT - 124

		9	6			1		
			7	8			4	
3					4			
						6		5
9		2				4		7
7		3						
			4					9
	8			5	9			
		6			1	5		

EXPERT - 125

		3			7	4		
			9	3				
1		9	8			3		5
5			1		3	9	8	
	6						5	
	9	1	7		5			4
4		2			8	5		3
				1	2			
		6	3			1		

EXPERT - 126

			1			7	4	
	4		7					
1		7					5	
5			4		1		8	2
	7				8			3
4			9		3		6	7
6		5					9	
	8		5					
			8			6	2	

EXPERT - 127

				8			2	
	4		2		6			
				3		4		9
		5			9	7		
9	3			2			8	4
		7			3	2		
				4		8		2
	5		6		2			
				9			1	

EXPERT - 128

7		5	8					
4					7			9
							7	8
		9		1	3			7
1			9		6	8		
		2		7	8			5
							9	2
6					1			4
9		8	6					

EXPERT - 129

		9				6		3
		3	8			2	5	
				3				4
	7		3	8				
	6			7			9	
				6	5		1	
1				9				
	5	8			2	9		
7		2				4		

EXPERT - 130

		5	1					
9				2		3	7	
7		8	6	5			2	
							1	8
			8	6	2			
8	5							
	9			8	7	4		1
	3	7		1				2
					6	7		

EXPERT - 131

			4			7	8	
6				8	1			
				3		4		
	1	2						7
	9		5		8		1	
5						2	9	
		9		5				
			6	9				2
	5	8			4			

EXPERT - 132

			4					
	3	7				1		5
				1	2	3		
9	6						2	
			3			4	6	1
3	4						7	
				8	5	6		
	7	5				9		4
			1					

EXPERT - 133

1	7			4	8		9	
	9		3				1	
6				9		4		8
							2	1
	3			1		8		
							5	3
4				5		3		2
	2		6				8	
9	1			3	2		6	

EXPERT - 134

	1	6						3
		2				4		
5	4				8	6		
				3	5			
	9	1						4
				9	2			
2	6				4	8		
		5				9		
	3	4						7

EXPERT - 135

				6				
	1		7		8		6	
9				5				3
		5				6		
		1	8	7	4	2		
2	8						1	4
		9				4		
		3	1		5	8		
1								6

EXPERT - 136

		3				8		
			1		8			
	1	8		3		4	6	
6								5
1			5	8	2			7
		5				1		
	8		4		6		7	
	5	1				9	4	
2								8

EXPERT - 137

3					1	8		4
			9					
				6		1	9	
	1	2			3	6		
9		4	5			3	1	
	5	3			9	4		
				9		5	4	
			2					
5					6	9		8

EXPERT - 138

						7	1	
8	5				2			9
		9		3			2	
	8		3	2		4		
2				8				6
		7		4	6		8	
	7			1		8		
4			2				5	1
	2	1						

EXPERT - 139

	5				7		4	8
9			5		6	7		
	1	6			5			7
3							1	
	4	9			8			6
5			7		3	2		
	7				1		5	3

EXPERT - 140

7								3
		5		3	6			7
		3	7				6	9
		9			7		4	
	5			1	9			
		7			5		2	
		2	9				7	4
		6		7	1			8
9								2

EXPERT - 141

	4	9	1		6	2	5	
			5		4			
			2		8			
		2				3		
9		6				4		5
	1		9		3		2	
	2						6	
		1				7		
6		7	4		2	5		8

EXPERT - 142

		3		8		6		
	2		5		3		8	
			2		4			
7	8						5	9
9				2				6
		2	7		9	4		
2				4				5
				3				
	1	6				8	4	

EXPERT - 143

					8	4	5	
		4		7		3		
		8	9					
2	6							9
	3		6		7			
5	4							1
		2	8					
		3		5		8		
					9	1	4	

EXPERT - 144

	9	1	6				7	2
				5		4		9
		3						6
9			2					
		4	1		8	7		
1			9					
		5						1
				2		8		7
	7	9	8				5	3

EXPERT - 145

	9	5	6					
6			5		8	9		
1		8			9			
	5					7		
2	1		8		5		3	9
		9					1	
			1			2		4
		2	9		4			5
					7	8	9	

EXPERT - 146

				1	2			
1	7		6					
	5	8	3					
		5				7	3	
7	2			5	1	9	8	
		9				5	4	
	4	2	8					
5	3		1					
				6	7			

EXPERT - 147

4		3		5		7		9
		9	7		6	3		
6	4	7				5	1	8
		5				2		
	2						4	
			5	1	9			
	6		8		2		3	

EXPERT - 148

	7	4	9					
				3	5	1		
3		2	4					
1						5		
		8	2		9	3		
		5						6
					8	4		3
		3	5	2				
					3	8	7	

EXPERT - 149

8			4			5	7	
	4	1	7		6			
				3				
5				1	7	4		9
9								6
1		8	3	4				2
				7				
			9		4	2	3	
	8	5			3			7

EXPERT - 150

2					1	6		7
				5				
8		4	7		3	9		
9		2				7		
	8			9			3	
		3				1		4
		8	2		6	3		5
				3				
3		1	5					8

EXPERT - 151

			4	5			3	9
				2		5		
						6	1	4
6			5			4		
9								5
		1			6			2
4	1	7						
		3		7				
2	6			8	3			

EXPERT - 152

		4	9		7	5		
	5						8	
7								9
1			5	7	3			6
5								8
2				6				3
4				5				2
	3						9	
9	1			8			7	5

EXPERT - 153

6	1	5	2		9	4	3	8
			1		3			
		4		6		9		
	9						8	
1	8						4	7
				9				
3		1		2		6		5
		2	7		6	8		

EXPERT - 154

		6	1				2	
4	7			2			8	
				7			5	9
		8			9		1	
		3				7		
	1		5			6		
2	8			1				
	6			8			7	1
	5				7	8		

EXPERT - 155

	4						2	
5			2		9			6
		6	3	2	4	8		
		4		8		7		
6		9	7		5	1		3
	1						6	
			8		6			

EXPERT - 156

7	8			2			3	4
		3				7		
	2		6		4		8	
		9	1		3	6		
				5				
	9	8		4		3	7	
	5						4	
2	3						9	1

EXPERT - 157

		3						
1						5	6	
5				6	4	2		
8			1		6		5	
	2						1	
	1		2		8			6
		2	6	4				9
	5	9						3
						8		

EXPERT - 158

5			6		1			
	7	3				6		1
1			3				9	8
			8	7				5
9				4	6			3
			9	1				6
3			7				5	4
	1	5				8		9
7			1		8			

EXPERT - 159

		4		5		7		
		2			9			
1	3						8	
			5			9	4	
			3	6	7			
			2			6	7	
3	9						2	
		6			8			
		1		7		4		

EXPERT - 160

7						6		
		6					9	
8			1	9	6			
					2			8
		4				1	6	9
					5			4
1			5	4	7			
		9					7	
2						5		

EXPERT - 161

	2				7	8		
		3		2				5
1			9				7	
7				6		5		
	5		1		3		4	
		1		8				7
	8				1			9
3				4		7		
		6	7				3	

EXPERT - 162

	5		3	2		1		6
6			8			9		
	1	3						
8				3		7		
			9		4			
		5		6				4
						4	5	
		1			3			8
4		6		1	8		3	

EXPERT - 163

	5		3	9				6
		7	6			5		
6	3					8		
		1		2			5	
7								4
		2		3			6	
1	8					9		
		5	4			7		
	7		5	1				2

EXPERT - 164

2					1	6		9
			2			7		
7	1	8			9	2		
5		3	6		4		7	
				9				
	4		7		8	3		2
		7	8			1	3	6
		1			7			
3		4	9					7

EXPERT - 165

4				1				3
	2	9				5	7	
8	1			7			9	4
7								8
			8	4	5			
	8			6			5	
			3		2			
		4				3		
		2		9		7		

EXPERT - 166

		1	3	8				
		5			7		8	
	6				9			
	5		6					9
1						7	5	4
	9		4					1
	3				4			
		8			1		2	
		9	7	2				

EXPERT - 167

		5	9					
		3		6	2	9	8	
				5			2	
8				3				4
		6				1		
1				7				2
				1			5	
		7		4	3	2	1	
		4	7					

EXPERT - 168

			6	4				
						3		4
4			5	7				
6		2		9			7	8
		3	8		7	5		
9	7			5		6		3
				8	4			9
7		1						
				2	5			

EXPERT - 169

	2	1						
	8		9	4			3	
		9			2	6		1
5					4	9	6	
				3				
	1	6	2					7
1		8	4			2		
	6			8	5		9	
						8	1	

EXPERT - 170

		2		9		3		
	4						1	6
7						9		
	8		7		9			
9		7	3					
	2		1		6			
3						4		
	1						7	2
		6		4		5		

EXPERT - 171

3					1	8		5
	8				3		9	
1		2				7		
4	5			3				
			1		2			
				7			1	3
		3				6		9
	9		2				8	
7		8	3					4

EXPERT - 172

	2						9	
		7		3		2		
8	3						1	4
5			4		3			9
	9	4				1	6	
			1		6			
			7		9			
		2	5	8	1	6		

EXPERT - 173

7								1
		5	9		8	3		
				5				
				8				
	9						3	
	3	7		2		4	9	
2	7		6	9	1		5	4
		6	5		7	1		
				4				

EXPERT - 174

		9		1		6		
	6		3		5		1	
	3	8	7		6	5	4	
9			4		8			5
4								3
	2						8	
		2	6	3	4	8		
			5		9			

EXPERT - 175

				3				
1	7						3	2
5			7		9			8
				9				
8								7
		3	6		7	5		
	5		3		2		4	
7		4	5		8	6		3
				6				

EXPERT - 176

		4	6		3			8
	1					3	9	
5			4					
			9	3			7	
1			7		4			2
	6			8	2			
					9			3
	5	3					2	
8			3		6	1		

EXPERT - 177

		9				2		
			2		7			
6	5		8		3		4	1
9								4
	7		4	6	5		1	
		4				5		
4			3		1			8
1				8				6

EXPERT - 178

					6			
5			8	7		9		
		6	1			4		
9		7	6				2	
	3	4				5		
1		5	3				6	
		2	7			1		
7			5	6		2		
					8			

EXPERT - 179

	5				7	4		8
						7	3	
	3		9				1	
			8	9				
		9	4		3	6		
				5	6			
	8				5		2	
	2	4						
9		3	1				4	

EXPERT - 180

	6	1				9	4	
	8	5				1	2	
7								3
		8	3		5	6		
	4						3	
			6		2			
	3		7		1		9	
				5				
5			4		3			1

EXPERT - 181

						6		
5			4		9			
7	9				8		2	
			6	1			9	
1		2				8		6
	6			2	7			
	3		9				5	2
			7		2			3
		8						

EXPERT - 182

			8		7			
			5		9			
8		4		6		9		5
5			1		3			7
7								3
		8				1		
	2						7	
	8			3			4	
9		5	6		4	8		2

EXPERT - 183

2	6		8		1	7		
	8							
		7	5			2		
	5			3				6
		4					8	
	2			7				3
		2	7			3		
	9							
5	7		2		9	1		

EXPERT - 184

6			5		7			3
		4	1	3		5		
	5						9	
5				9			4	2
	6		7		4		5	
4	8			5				1
	4						6	
		5		8	6	1		
2			9		1			5

EXPERT - 185

5			2			4		
				4	6		5	
				1		8		2
2	1							
3			9				7	1
4	5							
				3		9		5
				5	7		2	
6			8			3		

EXPERT - 186

		3					1	
5			2			9		
	4		8	1				5
			6		9	1	7	
		6				4		
	9	1	7		8			
3				9	2		5	
		9			6			7
	5					2		

EXPERT - 187

			4		1			
6			8		3			5
1								6
3		9	6		2	4		8
	7						5	
				7				
		5		8		3		
		8				5		
7	1		9		5		6	2

EXPERT - 188

		5						2
2	3				5			4
	8			6		5		
		7		8	6			1
					4			
		2		9	3			5
	2			5		7		
3	5				1			8
		9						6

EXPERT - 189

3			9	4	8			2
		8				3		
				7				
			4		5			
7		1				6		5
9				1				4
			7		6			
6	1						5	9
	3		2		1		4	

EXPERT - 190

					2		9	
5		7			6	4		
	2			4			1	
2	5		9		4			
		9				6		
			6		5		4	2
	1			3			6	
		2	4			3		7
	6		2					

EXPERT - 191

			4		6			
4	2						9	6
7				3				4
		8				9		
	7			8			3	
6			1		3			5
	6		8		5		2	
9		1		6		7		3

EXPERT - 192

					2			5
	7		5				6	
		5		9			8	
	2				8	6		
	1						7	
		8	1				9	
	9			8		7		
	3				7		4	
1			9					

EXPERT - 193

		4		6		5		
	6						9	
			2		1			
9			7		8			6
	8						7	
6		2	3		9	4		8
		1	4		2	6		
	4			3			5	

EXPERT - 194

4								2
1			4	5	8			9
	8		6	1	2		4	
			8		7			
	4						9	
8		1				4		6
		7	1	9	6	8		
6		8		2		9		5

EXPERT - 195

8		2				1		5
1	5						8	6
2	7		8		3		6	9
	1		4	2	6		5	
	2		3	8	5		1	
		6		9		3		

EXPERT - 196

6		3				8		
5	9		3					4
			5		9			
7			9		8	4		
8	3						1	6
		5	1		7			8
			8		5			
9					3		8	1
		8				3		2

EXPERT - 197

9					5	7		
7			3	6				4
			4			8	5	
							3	7
5			9	1	2			8
4	2							
	1	7			9			
2				7	3			9
		9	8					2

EXPERT - 198

4							3	
7		6			4	9		
				7		6	2	
	7		4	3				
		4	8		7	3		
				1	9		5	
	4	9		2				
		3	7			2		6
	1							3

EXPERT - 199

8				7	4			
	2			5			7	6
			8					3
9	6						8	
1			7		9	2		
7	3						4	
			2					4
	4			6			5	8
3				1	5			

EXPERT - 200

	9				2	6		
8			6		4		3	
1				5			2	
						2	5	1
		7			1			
						7	8	4
4				6			7	
6			1		3		9	
	8				9	5		

EXPERT - 201

	6		7		8	3		
		3			4			7
7				2			5	
1	4							5
		6				7		
3							8	9
	7			6				3
5			2			8		
		8	1		7		4	

EXPERT - 202

7		3			1		5	
				7		3		
2	5				9		6	
5			4		2	1		
		2	9		8			3
	7		8				2	4
		9		2				
	2		6			8		1

EXPERT - 203

						6		8
9				3	4		2	
4	8			6				
		1			8			7
8		5		1		4		3
		9			5			6
3	7			5				
5				2	7		6	
						5		4

EXPERT - 204

		1	7	6	8		9	
		9						6
6		8		9				5
			1					
		3		5			8	2
			9					
9		5		2				8
		6						4
		4	5	3	9		6	

EXPERT - 205

		1	8					
				9		6	4	
	2		7			9		
1							2	
2		3	6		9	4		1
	8							3
		2			4		8	
	4	6		3				
					7	5		

EXPERT - 206

5	7						4	8
	8						2	
				7				
	3			5			8	
4	1		8		6		9	3
				9				
		3				1		
7			9	6	4			2
	9		3		2		5	

EXPERT - 207

	7	8			6			2
			8					
		6		5				9
6		1					3	
8			4	1	5			
4		9					8	
		3		2				7
			5					
	4	5			7			3

EXPERT - 208

	3		9			4		
			6					3
		6		1	2	8		
	6	9		5				8
4								9
8				9		1	4	
		4	8	2		6		
2					4			
		5			9		2	

EXPERT - 209

		2						
4			3			9		7
		5		1		2		
	8			2	3	7		5
			7	4				6
	7			6	1	8		9
		9		7		5		
6			2			4		8
		7						

EXPERT - 210

2	5			4		1		
				6			4	
		7	8					6
6		1					9	
				8	9	3		
5		3					6	
		5	4					1
				7			5	
8	3			1		9		

EXPERT - 211

			2				3	
		8		6	3			4
					5	9		
2		5					8	
	6		4		8		5	
	9					2		1
		3	5					
6			1	9		5		
	1				4			

EXPERT - 212

8		6	4			2		
				9	2		8	
			3				7	
7	6				4			
	1	8		7		3	4	
			2				9	7
	4				3			
	9		5	2				
		2			9	7		1

EXPERT - 213

6				1				9
	2						7	
8			4		2			5
			6		7			
9	6						5	7
			5	9	1			
4								2
2	7	5				8	6	3
3			2	5	8			4

EXPERT - 214

3	8						2	6
	1	2		7		3	8	
	9						6	
8								4
		4	3	5	7	8		
2								8
9			8		5			3
	7			2			1	

EXPERT - 215

	9	5			4			
	1			6	5			
		8	3			2		
		1	9				2	
		9		3		4		
	3				7	5		
		7			1	9		
			2	9			5	
			5			8	4	

EXPERT - 216

	4	2	5			9		
				2				7
9						5		1
			8	7				9
6								5
5				1	9			
2		4						6
1				9				
		5			6	2	1	

EXPERT - 217

7				8	2			
			5				1	4
								7
2		7			3	9		
	8						2	5
9		5			8	3		
								9
			4				5	8
8				6	7			

EXPERT - 218

7				5	8			1
	4			7	6		3	
		3			1	6		
6	8	7						
4	3						1	6
						4	8	3
		8	5			2		
	1		8	6			9	
2			7	4				8

EXPERT - 219

		2		7				8
	3	5	9		8			
8					1			
5	9							3
	2	8				7	4	
7							2	9
			4					7
			5		9	4	6	
1				8		3		

EXPERT - 220

	8		4				5	
4				5				9
		2			9	7		
		8	9	1	6			5
	3		5		8		7	
5			2	7	3	4		
		4	6			9		
1				8				7
	2				1		4	

EXPERT - 221

8								7
7	5	9		3		6	8	4
			5		7			
4								6
2	7			4			5	1
	2		7		6		4	
		6	4	9	2	7		
5				1				9

EXPERT - 222

		4				1		
	1		3		5		8	
			1	7	6			
1		6		3		8		7
	2	5	7		8	3	6	
				2				
		9				4		
			9		2			
5	7						1	9

EXPERT - 223

		9				6		8
3					9			7
			1			5		
	4			1		8		6
	6	2	9				4	
	7			3		9		2
			7			4		
5					6			1
		7				2		5

EXPERT - 224

				7		3	9	
					4			6
						7	1	4
		4			3	9		1
	1			6			2	
8		6	2			4		
5	6	8						
1			9					
	2	7		1				

EXPERT - 225

7								
			4	6	1		2	
		2	9				3	4
2		6	3			7		
		5			2	6		1
4	3				5	9		
	8		2	9	4			
								8

EXPERT - 226

3			4	9	1			8
		5				3		
9								7
	9	6	5	4	2	7	8	
				8				
	3		7		9		4	
2	7			5			3	6
5				3				1

EXPERT - 227

				7				
	5		1		8		6	
4	8		6		3		1	2
2		8		1		7		3
1								8
7		4				3		9
	3	9	4		2	5	7	

EXPERT - 228

	5			6			2	
	7						6	
3			2		7			4
			6		9			
		2		8		5		
7								8
	6			1			3	
9				7				6
5		4		2		9		7

EXPERT - 229

			3					8
					5		7	2
		1	7			3		
4				1		8		
		8	4		3	1		
		9		6				5
		4			2	5		
9	2		5					
6					8			

EXPERT - 230

	7	5	3		2	8	1	
	8						9	
6	2			1			5	7
			9		4			
	1						8	
8	6	4				9	3	2
		6		3		1		
	5		1		6		4	

EXPERT - 231

8			9				1	4
1					4	2		
	9	4	6			8		
	5					4		8
3		7					6	
		5			7	1	3	
		8	2					5
9	1				5			2

EXPERT - 232

		6	2				8	
8			6			7		
						5		
		7		8	3			9
	1		5		7		2	
5			4	6		1		
		3						
		2			6			4
	9				5	6		

EXPERT - 233

3			6	8		7		2
		2		3		8		
6	8						3	
			9		4			8
8	3			7			4	5
5			3		8			
	2						6	7
		3		9		5		
4		8		6	5			3

EXPERT - 234

6			1	7	4			3
				2				
	2			8			9	
9		1				8		7
			9		7			
		5		1		9		
	5	9				4	3	
		2				6		
			8	4	5			

EXPERT - 235

	6	3			7	2		4
							5	
			5		4		7	
		1		4	5			
	4						2	
		9		3	2			
			9		6		4	
							8	
	8	2			1	9		7

EXPERT - 236

	3	2	5					9
			8		1	4	2	
	4							7
					8		7	
7	8			2				
					5		1	
	2							6
			4		9	2	5	
	9	6	2					1

EXPERT - 237

	1	4						
		5		9	6	8		
	6		2					7
			6		9	1		8
				2			3	
			3		4	7		5
	8		4					2
		2		3	7	4		
	7	9						

EXPERT - 238

						4		
		8		3			9	
			7		8		2	
	4			9				2
3	6			5		9	8	
	8			7				5
			6		9		5	
		5		1			3	
						1		

EXPERT - 239

	6			2			3	
			3	7	8			
2		3	4		6	5		7
4		6	8		5	7		1
7								5
	5						4	
3	8		7		4		5	2
		1		8		4		

EXPERT - 240

7					1	9		2
	4			2	3		6	
6								
4	3		2		8			
	9						2	
			4		5		1	3
								7
	7		8	5			3	
2		9	7					5

EXPERT - 241

	4		7		3			
6		8					3	1
7				6				
1	3	7			6	4		
				3	5			
4	9	5			1	6		
2				1				
3		9					6	4
	1		3		4			

EXPERT - 242

8								4
2		3		9				
		5				7	2	
			8		6	1		
	7	8				3	5	
		9	5		4			
	8	1				6		
				1		4		3
7								2

EXPERT - 243

					8		1	5
			2	1		4	8	
		8	3		4			7
3	9							2
		1					7	
5	7							1
		3	7		5			6
			8	6		5	4	
					3		2	8

EXPERT - 244

				2				
	4		7		1		5	
6		5		4		7		3
4	1			8			3	5
		6				1		
8		3				2		6
7	6			5			8	9
			9	6	3			

EXPERT - 245

			3	9	7			
	3			4			7	
1				5				4
	1			8			4	
		3	4		6	7		
2		4		3		9		6
8		1				3		7
6	4						9	8
		5				4		

EXPERT - 246

		9	5		8	3		
	4			2			7	
			1		3			
1		4				2		7
2								1
	3						6	
7		3	8		4	1		9
9	5						2	4

EXPERT - 247

9	8			3		2		4
		3				8		7
7	2				8		6	
		8		6				
4			3		9			5
				1		4		
	1		4				8	2
2		9				3		
8		7		2			4	6

EXPERT - 248

	1				5	9		
3		9		4		2		
	2			7	1			8
	4							9
7			1			6		
	8							1
	5			1	6			7
1		4		3		8		
	3				2	1		

EXPERT - 249

5				2				8
3	9		4		5		7	6
			7		3			
4		7				9		1
	3						8	
	1		9		4		3	
1								3
9	2		3		8		6	7
				4				

EXPERT - 250

		2			1			
		4		6		1		
	9			7			3	6
9				2				
	1	7	9		6	2	8	
				4				7
4	8			9			2	
		6		1		7		
			2			4		

EXPERT - 251

	1			7				9
	2				4	5		
		3	9	8			2	
		8	4					7
				5				
1					7	2		
	6			9	2	4		
		1	5				3	
7				4			6	

EXPERT - 252

				3	6			
		7				4		
	6		5		4		3	
2		6	8		9	7		
1				5				9
		8	1		7	5		2
	1		4		2		7	
		2				9		
			6	8				

EXPERT - 253

4							7	9
	5		2					
	8				4	5		
3	6			2				1
	7			5			9	
9				7			6	8
		6	8				2	
					2		3	
1	2							6

EXPERT - 254

1	9					6		
	8					3	9	
				9				1
		4	5	7			3	
3			8			7		
		6	2	3			4	
				1				9
	3					5	8	
2	6					1		

EXPERT - 255

	2		9	8				
		8				3		6
3				6				
4		5				7	1	
9				7			8	
8		3				4	9	
7				5				
		1				8		4
	8		2	4				

EXPERT - 256

9				2				5
		6	4			8		
	4				6		3	
		4		6			9	
3			2		9			4
	9			4		5		
	7		3				1	
		5			8	7		
4				7				8

EXPERT - 257

	7							
1					4	7	6	
	4		7			1		2
	8	7		6				
		4	2	5	7	8		
				4		6	2	
4		9			6		1	
	2	1	3					6
							3	

EXPERT - 258

6						2	5	
	2		4		8			7
			5				3	1
	9							
		7	2	1	4	3		
							7	
1	8				6			
2			3		1		4	
	4	9						3

EXPERT - 259

8	4		2				5	6
		2		6	7			
				5			2	
	3						8	2
		6	1			7		
	7						6	1
				8			4	
		8		7	3			
4	9		6				7	5

EXPERT - 260

			6	4	9			
7	6			5		9	2	
4								1
	4		8			1		9
		9		3		6		
6		1			7		3	
5								3
	1	4		7			8	5
			9	8	5			

EXPERT - 261

	5		4		3		7	
	8						2	
4			5		2			3
		1		2		3		
6				4				2
		9				1		
7	1						3	6
3				1				4
			6		7			

EXPERT - 262

				8	2			1
	6	8						
					6	8	7	4
7		6	2				8	
		2		7		1		
	9				1	3		7
6	8	3	5					
						7	6	
9			6	3				

EXPERT - 263

	8	4						
	9					1	6	
		6			2			8
		2	7			3	9	4
			2		5			
6	4	7			9	2		
3			1			4		
	1	9					2	
						8	1	

EXPERT - 264

6			3					
9			4	2	7	6		
4	8	2						
1		9			4			
				7				
			9			1		7
						9	2	5
		5	2	6	9			3
					3			8

EXPERT - 265

		6	4			9		
							7	
	1	4	8				3	6
		1		9				
	3						4	
		7		6				
	4	8	5				6	3
							2	
		3	2			1		

EXPERT - 266

	9			5			1	
5			1		3			9
		3	8		7	6		
2	8						6	4
7		5				9		1
9	5			1			2	6
			2	6	9			
		2		7		4		

EXPERT - 267

2	7	3	6			9		
	4			2				6
					9			
		9		5		2		7
	2						6	
6		7		3		5		
			5					
3				8			4	
		2			4	1	7	5

EXPERT - 268

	9							
				8	6	1		
6				5			8	9
9		6	5			4		
3	7				9			1
1		4	3			2		
7				2			3	4
				4	3	5		
	4							

EXPERT - 269

			1		5		4	
2		9		6	7			
								6
	5	1			4	9		3
3								2
	2	6			8	4		1
								4
8		5		4	6			
			5		2		3	

EXPERT - 270

	5	6	2		9	3	7	
	9		5		7		1	
		7				5		
5			8		6			3
	2	4				9	8	
	6	1		7		2	3	
	3						9	
			3	2	1			

EXPERT - 271

			3	9		2		
	9							5
3	1				6	9		
4	8							9
				3				
5							7	2
		3	7				4	8
6							2	
		5		4	3			

EXPERT - 272

8			1					9
		5		2		6		
	6		7	3			1	
			8		2	3		7
	8	4				1	2	
2		6	4		7			
	5			8	4		6	
		7		5		4		
3					1			2

EXPERT - 273

					2	6		
7			9		1			2
	5			4				
1				9	4			6
	8						4	
3			8	5				9
				7			2	
4			6		9			7
		3	2					

EXPERT - 274

		8			6	7	3	
	2			7				8
		5						
8		7	6	1	4			
	9			5			8	
			3	8	9	5		7
						4		
1				6			2	
	6	2	7			8		

EXPERT - 275

5								3
3		8				9		4
	4	1				8	5	
			2	5	9			
		9	3		7	6		
8				4				9
	8	4				2	3	
	9						7	
			4	2	6			

EXPERT - 276

	6	4				3		
7								2
	2		5	1			4	
		7		8				5
	8		4		7		9	
9				3		1		
	9			2	5		7	
2								4
		6				2	5	

EXPERT - 277

				7				
1			4		6			9
		6				2		
				5				
	7	5		3		4	9	
3				9				2
	1	7	5		2	9	3	
4		2	3	8	1	7		5

EXPERT - 278

1			6	4				7
		9		3		2		
	4				2		5	
		4						3
9	3						2	8
2						5		
	7		9				3	
		5		7		6		
8				1	6			9

EXPERT - 279

			9					8
	3	6	8					
	9			6		1		
3			4				7	5
		2				3		
6	7				2			4
		4		7			5	
					5	7	4	
9					4			

EXPERT - 280

				5		9		1
3					9	2		
		7			2			
		9		6	1		5	
4			3					
		3		7	5		1	
		2			7			
9					3	5		
				2		8		6

EXPERT - 281

4			3	1				
	3					5		
		1		8	5			3
				5			2	4
	7						9	
9	6			4				
8			5	9		3		
		6					5	
				2	4			6

EXPERT - 282

8								4
			1	7	4			
	7		3		5		1	
				6				
		5	9	2	7	6		
	9						5	
				1				
	3	2				7	4	
		9	4		8	1		

EXPERT - 283

				1			8	
7			6		4		2	1
		9		2	8			
9	3						7	6
	2					8		
5	6						1	2
		4		6	7			
6			8		1		4	7
				3			6	

EXPERT - 284

		3			1		6	
9				3		5		
			8		5	9		
1				7			3	
			3		6			
	2			5				9
		1	7		8			
		5		4				8
	7		5			6		

EXPERT - 285

					2			
2							6	7
7	4			6		2		5
	6	2		3				
8		1				6		3
				5		9	1	
9		7		4			2	8
4	3							6
			5					

EXPERT - 286

	7	1		3	6			
	9		1				7	
4			7					1
		9					4	
			8	6	7			
	2					1		
8					5			2
	3				1		5	
			2	7		8	3	

EXPERT - 287

6		5				2		8
4	9						6	7
		6	3		2	9		
2		7		9		3		4
		3		7		8		
			9	2	4			
3			8		1			9

EXPERT - 288

5				4				7
	6				5	2	3	
	9	4				8		
	2		7		4			
8				5				6
			8		2		7	
		2				6	4	
	4	7	5				8	
3				2				5

EXPERT - 289

5					7			2
	8				3	7	5	
	1	7	6		5	8		
9	2	1				5		
		6				4	2	8
		5	2		6	3	8	
	9	2	8				6	
1			3					7

EXPERT - 290

					6			
			5	9		3		
	2	8	4		3	6		
2		4	6		9	5	1	
	3						9	
	6	5	1		2	7		8
		2	9		7	8	6	
		7		6	5			
			2					

EXPERT - 291

2		8	4	3				
	5	4			1			
3	1				6			
	8			6	3		7	
5				9				2
	3		1	7			9	
			3				8	6
			7			2	1	
				1	9	3		7

EXPERT - 292

	7	9		3		1	2	
3								9
		5				8		
		8	6		1	4		
4			2	9	5			1
		7				5		
		4		6		3		
2				8				7
			7		9			

EXPERT - 293

	8	5		7		6	2	
	7		5		6		9	
		9				3		
	1			2			6	
5			1		7			4
		7	2		9	8		
9				8				6
	4						5	

EXPERT - 294

		4				2		
1								7
		8	5	2	4	9		
	8			4			9	
	9		1		6		8	
6			8		7			4
2	6	9				8	4	5
8								3

EXPERT - 295

		7		2				5
		3	8		1			
	4				6	9		
		2					3	7
5			9			6		8
		6					5	9
	2				7	3		
		9	2		5			
		8		3				1

EXPERT - 296

1	9			6		4	2	
							9	5
						7		1
	6		7		8	1		
				2				
		2	1		5		4	
2		6						
5	7							
	4	1		7			5	8

EXPERT - 297

9					2		4	7
2					6			
		7	4	1		2		
8	1					9		
		9				8		
		5					2	4
		8		3	7	4		
			9					6
5	3		2					9

EXPERT - 298

	2		4		9		7	
	5						8	
7				8				9
	7	8	9		3	5	6	
6								2
			2		1			
5		3				1		7
	6						5	
		7		2		9		

EXPERT - 299

	3			6			4	
7								2
	6	1				3	7	
8				3				9
	2		9		6		8	
	1	5		7		4	3	
		8				9		
			5		4			

EXPERT - 300

		8				9		
7								1
	2	1	8		9	5	4	
		2	1		5	8		
	9						1	
		3		9		2		
		7				4		
	3		5		1		2	
		5		3		1		

EXPERT - 301

				1				2
		6				1		
1	8			2		6		
6					3	2		7
	9	7			6			4
2					8	5		6
4	7			8		3		
		2				7		
				6				5

EXPERT - 302

				8				
7	1						9	3
			7		9			
2			8		5			1
		7	6		1	4		
		6				5		
		1		7		3		
9	5		4		2		7	8

EXPERT - 303

4	5		3					6
	1						5	3
		2			6	9		
		7		3				5
			8		4			
2				7		6		
		4	6			3		
5	2						6	
6					7		8	1

EXPERT - 304

5	7				4		2	
							9	
			1	9	5	4		
3				8		9		
1	5	6						8
8				4		7		
			2	7	6	3		
							6	
6	3				8		7	

EXPERT - 305

4			1					6
7		8				5		
	6			2				4
		6			7		3	
	8					6		
		5			3		4	
	7			3				2
2		1				9		
8			6					1

EXPERT - 306

		7	1		3	2	6	
3					9			
6			7	4				9
8	6					4		3
		2				6		
1		3					2	8
2				7	8			5
			3					6
	3	9	6		4	8		

EXPERT - 307

					7			1
		2				9	6	
6	5						3	
	6			7	8			5
8			9		6			2
2			1	3			4	
	7						2	9
	2	1				6		
5			3					

EXPERT - 308

			9	8	4			
7								8
	3			7			4	
	5			9			8	
9	1		8		5		2	4
		7				6		
			4	2	1			
6								2
	4						5	

EXPERT - 309

2	5				8			
		1	7				5	
		9	3			8		7
		2						
	9	5		1		3	4	
						6		
9		4			1	7		
	2				6	1		
			8				3	2

EXPERT - 310

		8			2			7
4				3		2		
				1	9	4		
8	4							
	3	5		4		8	7	
							4	9
		9	1	8				
		7		2				4
5			3			1		

EXPERT - 311

		7			8	4	1	
2								
	6				1			9
6			1	7				3
		2	6		4	1		
4			8	2				7
	8				2			4
9								
		4			6	9	7	

EXPERT - 312

		8		6		5		
				7				
		2	1		3	9		
		4		5		2		
	6		2		4		8	
6								5
	1	3	8		5	7	2	
		5				8		

EXPERT - 313

			8		5			
3				2			9	8
	1			9			5	
4	5					6		
	3	9					8	
6	8					2		
	4			8			7	
2				3			6	5
			7		6			

EXPERT - 314

	2	7					1	
3						8		
				2	5			9
7		4	5		8			
		8	6			4		
6		2	9		1			
				5	4			6
4						9		
	6	5					8	

EXPERT - 315

2			6		3			7
	1						6	
8			4		2			9
			3	5	9			
				6				
	6		7		1		4	
	3		8		7		1	
4								6
7								2

EXPERT - 316

5	2						4	8
		4	6		9	1		
		1				7		
2								9
			5	3	6			
	7						8	
			1		7			
				9				
	6	2	4		8	9	1	

EXPERT - 317

		8		4				
3	1	6						4
2			7				3	
	5			1	8	2		
			6					
	2			3	4	7		
7			4				9	
9	3	2						5
		4		5				

EXPERT - 318

2		6	4		1	5		9
7	3						1	4
		4		6		1		
		3	9		5	2		
9								6
		2		9		3		
		5	6		7	4		

EXPERT - 319

		2	6		8	3		
		1				4		
7	6						5	2
	1		7	4	2		6	
2				9				7
	5	4				7	2	
			3	2	4			
		6				1		

EXPERT - 320

						5		1
4		1		8				7
	7	3			2			
1							9	
	3	5	6		9	8	1	
	9							5
			2			3	8	
5				3		1		6
3		6						

EXPERT - 321

					6	1		
		5		2			7	
9			3					5
4	8		1		2			
			9	3		7		
3	1		7		5			
8			2					3
		6		9			1	
					1	8		

EXPERT - 322

4					6			
				5			8	
	5					4		9
8	7		2			5		4
	2		4				9	6
9	6		3			8		1
	4					9		2
				4			1	
6					8			

EXPERT - 323

5			3	9	8			2
2		1		4		7		9
		6	1		2	4		
7								5
	2	8				3	9	
1			7		4			3
9	4			5			1	7

EXPERT - 324

		6	2		7	5		
7			3		5			9
5								6
		1	6	9	8	3		
		3		1		4		
	5			7			1	
	9		4		1		5	
		7				8		

EXPERT - 325

8								
4		7			9		1	2
	5	3						6
6	8		2					
		5	8		1	2		
					4		8	9
7						6	4	
2	1		4			7		3
								8

EXPERT - 326

	7						8	
		2	8		3	5		
				4				
4		8	2		6	7		5
	5	7				6	2	
				9				
7			6		8			4
		6				3		
			5	2	1			

EXPERT - 327

	5		7	4	6			1
2		8	1		3			
		1						3
	6		9			4		
			8		7			
		2			4		5	
5						1		
			5		9	8		7
8			4	3	1		9	

EXPERT - 328

4			6			7		2
	8		1	4			9	
6								
			7		1		8	6
	3			2			7	
8	9		4		6			
								5
	5			1	4		6	
3		4			8			9

EXPERT - 329

3					2			1
	5			8	9	7	4	
	1		3					
4	7					1		
	8						9	
		1					6	2
					3		2	
	4	3	2	6			7	
8			5					6

EXPERT - 330

8			7					
		1			2	9		8
	5	3			1			
3							9	
	4	5	9		8	6	2	
	8							7
			2			3	5	
5		8	3			4		
					4			6

EXPERT - 331

	6						1	
1			7		6			4
3		4	5	8	1	6		2
			9		3			
	9						7	
				5				
	8	3	4		9	7	2	
				1				
6	1						4	5

EXPERT - 332

		6						
		8			5			
3	7		8			4		5
8				2				4
	3	4				2	8	
1				4				9
5		2			1		4	7
			9			3		
						6		

EXPERT - 333

				4				
5		2	3		6	7		9
8		7	2		5	6		1
				2				
			1		9			
9		8		3		1		6
6		4				3		2
				6				
2		9				8		4

EXPERT - 334

4						6		
	2				9		1	4
3	7				1			
		2				8		1
			8		7			
6		7				3		
			1				2	3
2	4		6				5	
		5						8

EXPERT - 335

	2		7	8				
3			5		9		7	
		7		1			8	
		5				1		
	1							9
		2				6		
		9		4			1	
4			3		7		2	
	5		2	9				

EXPERT - 336

		3		6	7			
					4	8		
	2		8	3				7
2	1					4		
9		7				3		2
		4					6	5
1				7	3		4	
		5	4					
			1	9		5		

EXPERT - 337

						5		
			7	3	5		8	
	3			9	8			1
9	1		4				2	
								7
6	5		2				9	
	4			1	6			9
			5	4	7		1	
						3		

EXPERT - 338

	6						4	5
3			4	1	2			
		9				3		
7		6	2	5				
2					7	5	9	
8		4	1	9				
		7				2		
6			5	2	4			
	2						6	4

EXPERT - 339

	6				2			
	2			8	6	1	4	7
	1		7	3				
4	5					2		
	9	8				3	7	
		7					6	4
				1	8		5	
5	7	1	4	2			8	
			6				1	

EXPERT - 340

		6	1				9	
	4		7		2		1	
				9	5			6
		8			9		2	
7				1				8
	6		8			5		
2			6	3				
	8		9		1		6	
	1				4	3		

EXPERT - 341

				8				
	8	3	6		7	1	2	
7								4
				5				
		1				6		
			7	3	1			
1	5		8		9		3	6
8		2	5		3	9		1
		9				7		

EXPERT - 342

	7			3		8	9	
					1		2	
		4	8					5
		3		4	2			
4								9
			9	8		5		
1					8	9		
	6		1					
	5	2		7			4	

EXPERT - 343

		9		5		1		
			7		9			
7		2				9		4
		4				6		
	9	8				3	7	
			3	1	2			
		1	5	4	3	2		
2				9				1
			6		1			

EXPERT - 344

2		5			3			7
		6				4		3
3			2			1		
	5	2		8				
	1			9			4	
				1		2	7	
		1			9			4
5		9				7		
4			7			8		5

EXPERT - 345

	8			3	7			6
7			8				4	
		9				7		
	1							8
		2	1	5	8	3		
	9							4
		1				8		
6			5				3	
	3			7	9			1

EXPERT - 346

	5				1			
		9	8		4			3
	8	1	6	3				
				4		9	3	
4	6							
				7		2	6	
	4	5	9	6				
		6	4		2			5
	2				3			

EXPERT - 347

	3				8		7	4
		7			9	6		
							8	9
7	6			5				
2	1	3		6	7			
8	4			1				
							5	2
		4			5	8		
	9				6		4	7

EXPERT - 348

7	1		2					3
		5	1				4	9
			5		6		1	
5		3						
	7			1			8	
						3		7
	5		6		4			
2	6				7	4		
8					1		3	6

EXPERT - 349

		9	4	2	5	7		
	4						8	
				9				
5								8
	2		7		8		3	
	8	3				5	9	
		8		1		4		
			9		6			
		7	3		2	1		

EXPERT - 350

	2	7				5	9	
				9				
	6	9	3		8	1	7	
	7						8	
			4		1			
		2	9		7	3		
6			1		3			8
2				7				6

EXPERT - 351

				3				
		4	1		8	2		
		9	5		7	8		
		5				1		
	4	6		9		3	7	
1			4		5			8
4	6						2	3
		8		1		6		

EXPERT - 352

				1		7	3	
		3			6			
1			8	3				
6	1					2		
		9	4		1	8		
		5					6	4
				8	3			6
			1			5		
	2	1		9				

EXPERT - 353

						7		4
			1		8		3	
		3		7		5	1	
8		4		5		9		
9		5		6				
6		1		8		3		
		6		2		8	4	
			6		4		7	
						1		5

EXPERT - 354

	7			8			3	
1	8						7	2
		2	9		7	4		
			6		9			
5		1				8		4
6	4		5		1		8	9
			3		6			
		9				1		

EXPERT - 355

			9		6			
		9				8		
3		1				9		4
			5	1	9			
	3	5	6	2	7	4	9	
				3				
1	2			5			7	9
	9		7		4		8	
	7						4	

EXPERT - 356

1								
5	2			1				
	9	7	8	3	2			
				8	3		4	6
		3	4			9		5
				7	6		8	2
	6	4	3	5	8			
3	7			6				
8								

EXPERT - 357

8		9				3	6	4
	6		8					
		2		9				
	4			1			7	
5		1		8		2		
	7			3			5	
		5		2				
	3		5					
9		7				6	1	5

EXPERT - 358

7	6		8					
		5				2		
				5	1		3	
		6		1			9	
	7	2		9		4	8	
	4			7		3		
	3		7	8				
		4				6		
					9		2	3

EXPERT - 359

	2	6				7	9	
					3			
		9			7	8		3
					1			4
		4	8	7	2	1		
1			3					
6		1	5			9		
			6					
	8	2				3	5	

EXPERT - 360

	3					1		
			8		7			
	1	7		3		6	2	
						9	8	6
3	6			2		4		5
						3	1	2
	7	1		5		2	6	
			1		6			
	4					8		

EXPERT - 361

	1				6	9		7
		9				3	6	4
			3					
5				3			2	
		7	8		4			6
6				5			4	
			5					
		8				6	1	5
	7				8	4		9

EXPERT - 362

	6	8	7					1
					5	6		3
9								
	4	9	2	8				5
	1			9			4	
8				4	3	9	1	
								4
1		7	4					
3					2	1	5	

EXPERT - 363

4		5						
		3		6	7			
9							1	
	5				1		3	4
8		9	5				6	7
	4				6		9	8
1							7	
		7		3	8			
5		4						

EXPERT - 364

1	9		5		7			
	2				8			7
8					3		4	
	8					1		
		6		4		8		
		5					9	
	4		3					8
6			8				5	
			2		6		3	4

EXPERT - 365

9		7						8
	4				8		3	
			2	3				4
	7		1		6	5		
		5		4		9		
		6	5		3		4	
7				1	9			
	1		6				7	
6						3		9

EXPERT - 366

6	2	1						4
		8	1	6		2		
4					2			
	6	7					4	
1								3
	9					7	2	
			5					2
		6		1	8	5		
9						4	7	1

EXPERT - 367

1			4	6	3			7
		9	8					
				1			8	
9							2	6
2		1		9		8		3
3	8							1
	1			7				
					2	3		
5			6	3	9			8

EXPERT - 368

	8		4					
				2	7		9	
		9	1			8	4	
3				8		6		5
			5			2		
1				3		9		7
		7	9			1	2	
				4	2		6	
	9		7					

EXPERT - 369

7		9	1					2
	3		6	4			7	5
5	9		2			4		
	4		7		5		9	
		7			3		6	8
3	1			7	6		5	
9					4	7		3

EXPERT - 370

		6				3		
5	2						7	6
			1		8			
	1		8	4	5		3	
2								5
			9		3			
7				3				2
			2		1			
	3			8			9	

EXPERT - 371

			1		9			
7		2				3		5
	5			2			4	
2								9
5		7				1		6
	9	1				4	2	
	3	6				7	1	
			7		1			
				8				

EXPERT - 372

6	3	5						
				1		7		5
1					4			
	7		8					1
	9		3			5	6	
	4		9					3
7					2			
				7		1		9
9	5	4						

EXPERT - 373

9		7						
4		3	1	7		9		
	1			3				
		6		2			8	
2		1				6		3
	3			4		5		
				8			7	
		8		5	7	2		1
						8		6

EXPERT - 374

	2	5				9	3	
		8	4		5	2		
6								5
		9				7		
3				7				1
	7		1		2		8	
2								4
		3		5		6		
			9		8			

EXPERT - 375

			6				2	1
						6		
3		9	2	8			7	
	8		3			7		5
	9						1	
1		5			7		8	
	5			2	6	3		7
		7						
6	2				8			

EXPERT - 376

4		3	8				9	
			6					
9	5	7					1	
		1	9					5
6			5		2			8
8					6	9		
	3					4	8	9
					3			
	9				5	2		3

EXPERT - 377

	3						2	
	1	6				3	5	
5		2	1		3	6		4
				9				
1		9	6		4	5		8
	9		5		1		3	
	6		3		8		1	
				2				

EXPERT - 378

		5	9		7	1		
7				4				6
1			5		2			7
9	5						8	3
	2			5			6	
		6	1		3	9		
	9						2	
			2		4			
	7						1	

EXPERT - 379

		6	9	4				8
	3	4		8	7			2
8						6	2	4
	2			3			8	
4						1	7	3
	4	2		7	6			9
		7	2	5				1

EXPERT - 380

				4			1	
9						3	7	
				5	3			
6		3	9			1		7
	8							
4		7	6			8		2
				6	2			
3						4	5	
				1			8	

EXPERT - 381

			5	4		8		3
							9	4
9	8				3			
	4				1	6	2	
1	7		6	2			4	
	9				7	1	8	
4	1				2			
							1	2
			1	6		7		8

EXPERT - 382

	6	1				4	8	
	4						5	
3				8				1
		9				2		
		4				1		
1				3				7
				9				
8	9		1		4		2	3
4			2		3			6

EXPERT - 383

			9			5	8	
9	7						1	
2		5			3	7		
		1		2				8
			4		8			
7				1		2		
		7	6			3		9
	9						6	4
	6	4			2			

EXPERT - 384

	7						5	
		2		8		4		
3				2				8
		1				5		
	3		7		9		1	
	2		8		6		3	
	6		3		5		2	
		5				6		
	4						9	

EXPERT - 385

2			6		3			8
			1	7	2			
	5	6		8		3	2	
	1	5		3		2	8	
9	2						7	3
4			7	1	9			6
		8				9		
			8		6			

EXPERT - 386

				1		2		
	5		3			8		9
			7		5		4	
1		6	4					8
			8		7			
5					6	9		4
	4		6		3			
2		8			9		6	
		3		8				

EXPERT - 387

		9				8		
			8				3	
	5		7		9	6		
4	3					2	6	
		8			1			
5	7					3	9	
	2		6		3	9		
			4				2	
		4				1		

EXPERT - 388

						9		7
			7		8		6	
	6				5	1	3	
		3						6
	4		6		3		9	
8						2		
	3	9	1				4	
	7		3		9			
1		2						

EXPERT - 389

	2			6				3
			9	2		4	1	
	4			7		6		
		6			8	9		
8	1		6					
		2			7	8		
	3			4		1		
			1	3		7	8	
	7			8				9

EXPERT - 390

					7	5		
	6	7					1	
4			6				2	
2			7	4	9	8		
			8		2			
		5	1	3	6			2
	8				1			7
	1					9	5	
		2	4					

EXPERT - 391

		2		6	8			
			9					
8			4		3	5		9
4						6		5
	2						4	
6		5						8
5		1	8		9			7
					5			
			2	7		1		

EXPERT - 392

	9	1	8		6	5	4	
8		2		3		9		1
9								4
7			1		5			3
		8				6		
		4	3		9	2		
				1				
2		5		8		7		9

EXPERT - 393

6			8		4			
	8	7				6		
	1		2	7		3		
			4		9		6	
		1				7		
			7		3		8	
	9		3	2		8		
	7	2				9		
4			1		7			

EXPERT - 394

		7					5	3
	8			6				7
		5	7			8		
	6	9	1				8	
					8	4		
	7	8	6				3	
		2	5			6		
	5			4				2
		6					7	4

EXPERT - 395

	3	6	7		1	2		
1								9
		2	6					
				4				8
8	4						2	3
7				6				
					6	3		
5								2
		9	2		8	7	4	

EXPERT - 396

		8			9			
	2				5	3	6	
	7			1				8
1	6							
		2		8		9		
							4	7
7				3			9	
	9	3	1				2	
			4			8		

EXPERT - 397

			1					
		2				1		6
8				4	9			3
	5		8				6	7
				7				
9	7				4		3	
3			7	8				4
1		8				5		
					5			

EXPERT - 398

				2	3	9		7
		9		4			2	
5						3		
	9					4	7	
1		3			9	5		
	6					1	9	
8						6		
		1		3			5	
				8	1	7		9

EXPERT - 399

			3		4			
	1						7	
		8	9		2	1		
8			1	2	3			6
	6	1				8	9	
	5			6			2	
7		9				6		2
	8						1	
		5		3		9		

EXPERT - 400

		1			2	5	4	
2	6		4				9	
8				1				2
4							1	
		9				2		
	3							6
6				7				9
	2				1		6	5
	9	4	6			3		

EXPERT - 401

			8	2	6			
4			9		1			3
2		3	6		4	7		8
		8		5		9		
7				6				9
		9		1		4		
3		2				8		7

EXPERT - 402

			6		3	1	7	
3		4				6	8	
								3
2		9		4				
	3				1			
5		7		6				
								1
9		8				7	3	
			5		2	4	6	

EXPERT - 403

		1	5		7	3		
	6	3				2	5	
	5						4	
6	3			7			8	2
		9		8		5		
		4		6		7		
				9				
			6		3			
7			8		1			3

EXPERT - 404

	3		9	7	1	6		4
				8				
					6	7		
		3			4	1	2	
	7	5						
		1			7	5	8	
					2	8		
				4				
	8		7	5	3	4		2

EXPERT - 405

							1	
3			4	5		8		
	9	6						4
	8			6	1		9	
		5		7		1		
	6		2	9			8	
8						7	5	
		4		3	2			8
	5							

EXPERT - 406

		8		9		1		
	2	7				3	4	
			1	2	3			
	3			8			1	
	8			4			9	
5		9				6		7
4				3				8
			6		4			
		3				7		

EXPERT - 407

5		7		8		2		
			4					
4	8				5	6		
		4	8		9			3
			3	6			1	
		2	7		1			6
7	4				8	5		
			5					
1		6		7		4		

EXPERT - 408

		8	7		4	5		
		1		9		2		
5	9						4	8
			3	6	2			
3				5				9
				8				
6			5		9			2
	1	9				4	6	

EXPERT - 409

8			3	4	6			7
5								2
		1				8		
			6		9			
	5	9				4	1	
		8				3		
			4		2			
2				3				4
		3		5		7		

EXPERT - 410

4	6						1	2
	9						6	
		2	1		7	4		
	4						9	
5		7				1		6
		6	3	7	8	9		
2			6		1			3
8			5	9	2			1

EXPERT - 411

		9				2		
	6						3	
2	7			8			9	6
	9		7		3		6	
8	3			4			7	9
	1	7	2		9	3	8	
			3		4			

EXPERT - 412

8		6				9		3
		3		1		8		
			3		8			
			6	7	4			
	9						8	
	7		8		5		1	
3								5
	6						2	
			2	5	7			

EXPERT - 413

					1	5		
	2	9	3				4	
4			8				6	
7				1		4	8	
			2		8			
	8	2		9				6
	3				2			1
	4				6	3	9	
		8	1					

EXPERT - 414

6							4	8
3				6	4	5		
	5		3					
	2			8		1		
	4		5		1		8	
		1		4			5	
					3		7	
		7	9	1				5
2	1							9

EXPERT - 415

		1	4		5	6	7	
5	7			1			2	
8			2					1
7						9		4
	9						1	
2		5						6
1					9			5
	3			2			8	7
	5	2	8		1	3		

EXPERT - 416

								9
		8			6		4	5
2			9			8		
	3			1		7		
	1			3			2	
		6		5			3	
		7			5			6
5	8		7			4		
3								

EXPERT - 417

	2	4		5		3		
8						6		5
	6				7		8	2
			1				2	
		6		3		9		
	5				2			
6	7		5				9	
9		5						3
		3		2		7	5	

EXPERT - 418

9	2		7					
	3							
	6	8		9	3		4	
	4	7	2			3		9
5			3			7		
	8	6	4			5		1
	7	4		5	2		9	
	1							
8	5		6					

EXPERT - 419

1		8	4		5	9		2
	3						8	
7								4
4			7		1			8
	8		9		2		5	
2		7		5		8		1
			1		4			
	1			2			6	

EXPERT - 420

	3				2	8		1
				8				3
		2				9		
4	2			5	8			
	5	6	7				3	
3	7			4	1			
		7				4		
				1				9
	1				4	5		6

EXPERT - 421

	2		9		3		1	
		8	1		4	9		
9				2				4
	3		8		7		6	
8								9
	4	1				8	2	
		4	6		9	2		
3								8
				5				

EXPERT - 422

8	1		4					3
		2	7		8		6	
6						9		
	8				4	3		
5				1				6
	6				3	7		
9						4		
		6	9		1		3	
1	5		3					9

EXPERT - 423

	5					3	6	
			7					4
2	8		3	4				
	6							8
9		8		2		6		7
1							3	
				8	1		2	6
4					2			
	2	6					7	

EXPERT - 424

6				9			4	7
7		9	6		2			
					5		2	
	4	3		1			6	
2			4		6			3
	6			8		2	7	
	9		5					
			9		1	7		5
1	5			2				9

EXPERT - 425

	1				4			
8	3	7		6	5	4		
5		9						3
	9							
6	2			5			3	7
							1	
1						7		9
		4	9	1		6	8	5
			7				4	

EXPERT - 426

6					7		8	5
3	7		9	8			6	
		5		3		9		
5							2	
	4	2				5	9	
	9							1
		8		9		2		
	6			5	3		1	9
9	3		2					6

EXPERT - 427

9					3			4
	8		5			7	9	
	5			4				
2				8			7	
		6	3		1	2		
	9			2				1
				9			6	
	4	9			7		1	
5			8					7

EXPERT - 428

3		8			2			4
	6					1	2	
	1		9	6				8
1				9		2		
		9	4		1	8		
		7		8				1
7				1	5		8	
	3	1					6	
8			3			7		5

EXPERT - 429

	5		8	4				
3			6				2	
4	9							5
1	2					9	8	
				7				
	4	5					1	6
9							4	1
	8				1			9
				8	3		6	

EXPERT - 430

	8						3	
		3				5		
5		9		3		1		6
				9				
			2	4	1			
1			8	7	3			4
8		2				3		7
		7				4		
3		4	7		5	8		1

EXPERT - 431

				4				
	5						9	
3		1				4		8
	8		4		9		7	
			8	2	5			
		3		1		5		
5			7		6			1
	6	7	3		1	8	5	

EXPERT - 432

5		6	1		8	9		4
			4	2	5			
7								8
	7			4			5	
			8		7			
6				3				7
		9	6		1	4		
2	8						1	9

EXPERT - 433

	9							5
			4			8		7
8		2	5			9		
	4			8				1
		3	2		5	7		
9				6			4	
		1			2	4		3
6		9			4			
4							5	

EXPERT - 434

	3			4			5	
2					6			
	4	6			8			7
	8	5	1					2
							8	6
	6	7	4					5
	5	2			4			8
6					7			
	7			5			4	

EXPERT - 435

3	7			2			9	5
	2	5		3		6	7	
1			7		6			4
7								3
	9		2		4		1	
9			3		5			8
	5			1			3	

EXPERT - 436

				8				
	1		3		7	6		
7	3		6	9	4			1
	9						1	
		6		1		4		
	8						6	
9			7	3	6		8	4
		5	8		1		3	
				5				

EXPERT - 437

			5		1			
	4	2		8	3		9	
					2		3	
1	6	8						3
	9						1	
5						6	8	7
	8		6					
	5		9	2		3	6	
			1		7			

EXPERT - 438

	2							
		6	8			1	7	
9					1			5
	5			4		6	1	
		9				5		
	8	4		5			2	
8			7					9
	1	7			2	4		
							3	

EXPERT - 439

3		8				9		
2			7					
				3	2	1		
	9		3				5	2
	7		9		4		3	
8	2				7		9	
		6	1	9				
					8			4
		7				5		9

EXPERT - 440

	9			7			3	
			1		9			
	7	2				1	4	
			4	9	8			
1			5		2			8
2	6			5			1	4
9	1						7	5
		4				8		

EXPERT - 441

		2				1		
1			4	5	6			8
		9	7	1	3	5		
		5		4		9		
	8	6				3	1	
				2				
	5						9	
	6	1				8	2	

EXPERT - 442

3			7				4	
		2	5			7		
					9	1		
	9				5			7
		8		1		4		
6			9				3	
		1	4					
		9			7	2		
	6				3			5

EXPERT - 443

6							4	
	4			5		9		
		3	4			5	6	
			2	8		1		6
2		9		1	4			
	7	5			8	4		
		4		9			5	
	6							7

EXPERT - 444

	1		2	4	5			
	2				6		8	1
5	9		6		4			3
1								8
2			1		9		4	5
7	4		5				1	
			4	6	3		9	

EXPERT - 445

1	8			3				
			6		4	8		
				1				3
	9				5	7		
		7	4		1	9		
		2	9				4	
3				2				
		5	1		8			
				9			6	4

EXPERT - 446

		8	7			6		
	3		6				1	
4				3	1			5
		5					8	2
		7				4		
8	2					5		
2			1	5				6
	5				6		4	
		6			4	1		

EXPERT - 447

	9	5				3		
2				8			6	
4				2	3	8		5
					4			7
		2	6					
					1			6
6				4	5	7		8
1				7			5	
	4	7				1		

EXPERT - 448

1			7	5				2
7	4	5	8					
				9				5
9	1						2	
4		6						3
5	3						4	
				8				1
8	7	3	1					
6			3	7				4

EXPERT - 449

			1	8	7			
	8	2	4		9	7	1	
2				3				7
		3				2		
9		7				1		8
			8	5	3			
4				7				2
		1				3		

EXPERT - 450

							3	1
3	8	1				6		
	9		6					5
		6	2					
9	1	8		6		3	4	2
					4	8		
6					5		9	
		5				7	2	3
1	2							

EXPERT - 451

5		2					7	
						9		8
		9			6			
4			8	6				1
8				5				3
3				1	4			5
			2			8		
7		6						
	5					3		7

EXPERT - 452

			9			4		7
			1					3
					7	6		9
1		2		6			3	8
		4		3		9		
8	3			1		5		4
5		8	2					
3					5			
2		7			1			

EXPERT - 453

		8					5	
4	5		7		1			
	6				4	9		
9								4
	7		2				3	
6								8
	4				6	3		
5	2		1		9			
		7					6	

EXPERT - 454

					7			
5	8				4			
9			5		3		7	1
	2							8
	4		8		9		6	
7							5	
3	5		9		8			6
			6				8	3
			7					

EXPERT - 455

6							1	5
			9	2			3	
						9	4	
4				9		5		
			2	4	5			
		2		3				7
	8	7						
	3			6	9			
5	9							3

EXPERT - 456

			2		1		8	
3	6						1	
		8	6	7		5		
2			3		5	7		8
		6				3		
5		3	8		7			6
		1		8	6	2		
	3						6	1
	9		1		3			

EXPERT - 457

6			5	1			9	
9	8				6		1	
		7		2				
		1				2		9
				7				
4		2				1		
				6		4		
	6		4				2	5
	4			3	5			8

EXPERT - 458

	8			7			2	
	1			8			9	
		2				3		
3			4	9	7			1
1			3		8			7
				5				
2	6			1			7	3
			2		5			
	4	1				8	5	

EXPERT - 459

	4				3			
3			4		8			
1		2		5		3		
	2	7				1		
8								3
		1				5	4	
		6		8		2		5
			7		2			1
			1				8	

EXPERT - 460

	7	8		6	3			
5								
	2					1	5	
			2	1	6			5
6								7
			3	7	8			1
	6					8	4	
3								
	5	9		2	4			

EXPERT - 461

9	6		7					8
		4			2	3		5
	3						1	
	4		5		1			3
				7				
1			3		8		9	
	8						3	
5		3	6			1		
6					9		7	4

EXPERT - 462

		4						2
			6		5			3
7	9						1	
9					2	3		
		3	4					
5					1	8		
1	8						3	
			2		8			7
		9						4

EXPERT - 463

	5		1					9
4	8							
		9		7	8			
		2	9		5		7	
	4					2	6	
		8	3		4		9	
		7		5	1			
8	1							
	9		7					1

EXPERT - 464

		4		6		1		
			4		8			
	7	2				8	3	
2	5	3		4		9	6	8
		9				7		
	6			3			4	
	4						5	
			7		6			
		8				2		

EXPERT - 465

8	6				2			
		3	6					
	2		9				3	
			7			9		5
	9	7		8				
			5			2		1
	7		8				5	
		4	3					
2	5				4			

EXPERT - 466

6		5		4		1		3
4								8
3	2						4	9
		4		7		9		
	3			1			7	
		6				3		
9			7		8			2
			4		3			
	5						3	

EXPERT - 467

	2		9		1	5		
			3			4		8
4	1			8				
8				1			6	5
		2	5		7	1		
6	5			3				4
				2			5	1
2		6			4			
		5	7		3		4	

EXPERT - 468

		4			7			
	8	9	1					
	5				9		6	8
			9		5	8		6
5								3
2		6	8		1			
4	1		7				5	
					3	4	7	
			4			3		

EXPERT - 469

1			6		5			3
		9	8		2	6		
	9			2			5	
6	1						7	8
		5				1		
5	8						4	9
	3			6			2	
			7		9			

EXPERT - 470

		8	2		4	7		
3		7		1		4		2
		2		3		5		
8				4				3
			6		1			
		6	1	5	7	9		
	8						6	
9				6				5

EXPERT - 471

8		1	4					5
			7			9		
	9		2	3				7
						7	3	1
		4				2		
1	3	2						
3				9	8		5	
		8			2			
6					7	8		3

EXPERT - 472

						8	6	
6	5		9					
	7	3				5		2
	6	1	7		9	4		
				1				
		9	3		8	1	5	
3		5				2	8	
					3		1	9
	1	6						

EXPERT - 473

9			5	2			8	
				1	3			
					8			1
5	1					3		
2		9				4		5
		3					7	8
1			4					
			6	5				
	4			8	7			6

EXPERT - 474

		5		7				8
6		8	2					5
7			3					
		1		6				9
4								2
5				1		6		
					7			6
8					3	5		1
9				4		7		

EXPERT - 475

				6			1	
		8				6		
	5		4	9	3	2		
			5				7	2
1					7			4
			8				5	3
	8		2	7	5	4		
		4				7		
				8			2	

EXPERT - 476

		5				4		
			1					5
9	8		6	7				1
	9				6			
2			7			6	5	
	3				4			
7	2		4	1				6
			5					8
		9				3		

EXPERT - 477

9	4						8	3
				9				
	1		2		8		6	
				8				
		9	1		3	8		
3								5
1				7				8
		2		5		3		
	5		3	6	4		9	

EXPERT - 478

		1				3		
3			4		2			7
	5		7		6		9	
6								8
		7	2		5	6		
9				6				4
	3			4			8	
	1		9		8		5	

EXPERT - 479

7	2			8			6	9
		9	6		5	7		
			5		2			
	5	3				8	9	
9	4		8		1		7	5
		5		6		2		
	7						5	
8								4

EXPERT - 480

	7		8		3		5	
			9		5			
		4				9		
	2	5				3	1	
	6			2			7	
	9	7				8	4	
	5			6			3	
3	1			9			2	7

EXPERT - 481

		2	9	5	6	7		
		7				4		
6								9
			8		5			
	7						1	
			3	9	1			
7		5		2		1		6
4	3						9	8
	1						7	

EXPERT - 482

1			6	4		8	5	9
3								
4		8			9	3		
		7		5				3
6			8		1			2
8				7		9		
		1	9			2		8
								1
7	4	3		2	8			5

EXPERT - 483

		1			9	4		8
2				1				3
		5		3			6	
			9			1		
			3	2	7			
		7			8			
	5			7		3		
8				9				7
4		3	6			2		

EXPERT - 484

7			9			5		8
	5		1		7			
		8						4
		2		1			4	
		7		6		3		
	3			5		7		
2						8		
			2		1		6	
9		6			8			7

EXPERT - 485

				4	7		2	
3	2	1	5		9		4	
			2				6	
9	3					8	1	
7								2
	1	5					3	7
	9				4			
	7		9		2	3	8	4
	5		8	3				

EXPERT - 486

		9				7		
6		7		3		1		5
		2	1	5	7	4		
		1				8		
7			3		5			9
9				8				7
			5		9			
8	9						4	1

EXPERT - 487

				9				
	3		1		7		8	
9	6						1	3
5	9						6	1
3	1			7			2	5
			8		4			
			6	5	3			
		6				8		

EXPERT - 488

5							3	7
	7	3			6			
9					5	4		
				4				5
	5		1		3		7	
6				9				
		7	2					4
			3			7	8	
4	6							9

EXPERT - 489

7	9				1		4	3
2		4	8		6			9
				9			6	
8	5						9	
		1				5		
	4						3	6
	7			8				
1			4		9	3		7
4	8		7				1	5

EXPERT - 490

	3	7			2		8	
8		5	6			2		
			3		8	5		
		1					3	
2	9						1	4
	6					9		
		2	8		6			
		8			1	3		9
	7		2			8	6	

EXPERT - 491

			5		4			8
	4			2	8		9	
		2		9			3	
					1			4
4		3				9	7	
					6			2
		1		7			5	
	3			8	5		4	
			2		3			7

EXPERT - 492

					8	5		
			4	7				
8			3			1	4	
		5	9		4		1	
	4	7		3				6
		2	7		6		3	
5			6			3	2	
			5	9				
					1	4		

EXPERT - 493

	5		2				4	
2					3			5
					4	6		3
9		5	7					8
				9				
6		7	4					2
					8	2		6
8					9			7
	3		1				5	

EXPERT - 494

5	6						8	3
8			6		7			5
	2						1	
			4	9	5			
3				8				4
		4				1		
1								9
2	8		5		6		4	7

EXPERT - 495

7		2		1		6		5
4	5			9			1	7
			5		7			
		7				3		
5								8
9	4						6	1
				6				
	1						4	
	9		4		3		8	

EXPERT - 496

							7	6
	1		8					
6		5						9
		7	5		2		6	
		2	7		4	3		
		8	6		3		1	
5		4						7
	9		2					
							4	8

EXPERT - 497

		4		8				5
		9	4					3
		6			2	8		
	7						1	9
3			6					
	9						6	7
		3			4	2		
		7	8					4
		1		2				8

EXPERT - 498

		6	7		3			
		3		9		8		
2	5	7			4			
				4				2
9	7		6	8				
				2				9
4	6	5			2			
		9		7		1		
		1	9		8			

EXPERT - 499

9								
		5	3			8	6	1
	4	1			6			
		8	1	7			5	2
	7		6					
		9	4	2			8	6
	9	7			5			
		4	2			5	9	7
2								

EXPERT - 500

			2		9			
	8	5						
				3			7	4
		9			4	2		
			1			3		8
		2			7	6		
				5			6	3
	1	7						
			4		8			

SUDOKU
SOLUTIONS
EASY - {1 -200}

EASY - 01

7	5	3	1	9	4	2	8	6
1	8	4	2	6	7	9	3	5
2	9	6	5	8	3	1	4	7
5	4	1	7	3	8	6	9	2
3	7	9	6	1	2	8	5	4
8	6	2	9	4	5	3	7	1
4	3	5	8	2	6	7	1	9
9	2	8	4	7	1	5	6	3
6	1	7	3	5	9	4	2	8

EASY - 02

4	5	2	6	9	7	1	3	8
6	1	8	5	4	3	7	2	9
7	9	3	8	2	1	4	5	6
3	2	9	1	5	6	8	4	7
1	6	5	4	7	8	2	9	3
8	7	4	2	3	9	5	6	1
5	3	7	9	8	4	6	1	2
2	8	6	3	1	5	9	7	4
9	4	1	7	6	2	3	8	5

EASY - 03

5	3	6	9	7	4	8	1	2
8	7	1	6	3	2	9	4	5
9	4	2	5	1	8	3	7	6
1	8	4	2	9	5	6	3	7
7	9	5	4	6	3	2	8	1
2	6	3	1	8	7	4	5	9
6	1	8	7	4	9	5	2	3
4	5	9	3	2	1	7	6	8
3	2	7	8	5	6	1	9	4

EASY - 04

8	6	2	9	4	3	5	7	1
4	1	7	2	6	5	8	9	3
9	5	3	1	8	7	6	2	4
2	7	8	3	5	6	1	4	9
1	9	4	8	7	2	3	6	5
5	3	6	4	1	9	7	8	2
3	8	9	6	2	1	4	5	7
6	2	5	7	3	4	9	1	8
7	4	1	5	9	8	2	3	6

EASY - 05

9	6	5	4	7	1	2	3	8
8	3	4	6	5	2	7	1	9
1	2	7	8	3	9	4	6	5
7	8	6	3	9	4	1	5	2
3	9	1	5	2	6	8	7	4
4	5	2	1	8	7	6	9	3
5	4	3	7	1	8	9	2	6
6	1	9	2	4	3	5	8	7
2	7	8	9	6	5	3	4	1

EASY - 06

2	7	9	5	1	3	4	8	
3	6	5	8	2	4	7	1	
4	1	8	6	9	7	2	5	
9	8	7	1	4	5	3	6	
1	3	4	2	7	6	5	9	
6	5	2	9	3	8	1	7	
8	4	6	3	5	1	9	2	
7	9	1	4	6	2	8	3	
5	2	3	7	8	9	6	4	

EASY - 07

8	6	4	5	7	3	2	9	1
3	7	1	9	2	6	8	4	5
2	5	9	8	1	4	6	7	3
7	1	2	4	6	8	5	3	9
5	9	6	7	3	1	4	2	8
4	8	3	2	9	5	1	6	7
9	2	5	1	4	7	3	8	6
1	3	7	6	8	2	9	5	4
6	4	8	3	5	9	7	1	2

EASY - 08

5	2	3	7	9	6	1	8	4
1	7	4	8	3	2	6	9	5
9	6	8	1	4	5	7	3	2
2	1	9	3	5	8	4	7	6
4	8	6	9	1	7	5	2	3
7	3	5	6	2	4	9	1	8
6	9	7	4	8	3	2	5	1
8	4	2	5	7	1	3	6	9
3	5	1	2	6	9	8	4	7

EASY - 09

1	2	6	3	4	7	5	9	8
9	4	7	8	6	5	1	2	3
3	8	5	2	9	1	4	6	7
7	9	2	1	5	4	3	8	6
4	5	3	6	7	8	2	1	9
8	6	1	9	2	3	7	5	4
2	1	8	7	3	6	9	4	5
5	7	9	4	8	2	6	3	1
6	3	4	5	1	9	8	7	2

EASY - 10

2	4	9	3	1	7	8	5	6
6	5	8	2	4	9	7	1	3
1	3	7	5	6	8	9	4	2
5	1	2	7	9	4	3	6	8
8	9	4	1	3	6	5	2	7
7	6	3	8	5	2	1	9	4
4	7	5	9	2	3	6	8	1
9	8	6	4	7	1	2	3	5
3	2	1	6	8	5	4	7	9

EASY - 11

9	7	6	8	5	1	3	4	2
4	1	8	2	3	7	9	5	6
3	2	5	6	4	9	8	7	1
7	3	2	1	9	6	5	8	4
5	8	1	4	7	2	6	9	3
6	9	4	3	8	5	2	1	7
1	5	3	7	2	8	4	6	9
2	6	9	5	1	4	7	3	8
8	4	7	9	6	3	1	2	5

EASY - 12

6	9	5	3	8	2	4	7	
2	1	4	7	9	5	8	3	
7	8	3	4	1	6	9	2	
5	7	1	6	4	8	3	9	
4	6	8	2	3	9	1	5	
3	2	9	1	5	7	6	4	
1	3	2	5	6	4	7	8	
9	5	6	8	7	3	2	1	
8	4	7	9	2	1	5	6	

EASY - 13

9	8	7	6	1	3	2	4	5
3	4	5	2	7	8	6	9	1
1	6	2	5	4	9	7	8	3
5	7	1	8	9	4	3	6	2
2	9	8	3	6	7	1	5	4
6	3	4	1	2	5	9	7	8
4	1	3	9	8	6	5	2	7
7	2	6	4	5	1	8	3	9
8	5	9	7	3	2	4	1	6

EASY - 14

1	3	4	7	2	5	6	8	9
5	2	8	6	9	3	4	7	1
9	7	6	4	8	1	2	5	3
3	5	2	9	1	7	8	4	6
7	6	9	8	3	4	5	1	2
8	4	1	2	5	6	9	3	7
4	8	3	1	6	2	7	9	5
2	1	7	5	4	9	3	6	8
6	9	5	3	7	8	1	2	4

EASY - 15

8	2	9	5	1	4	3	7	6
5	4	7	9	6	3	8	1	2
3	6	1	7	2	8	4	5	9
1	9	6	8	5	7	2	4	3
4	8	5	2	3	6	1	9	7
2	7	3	1	4	9	6	8	5
6	1	2	4	7	5	9	3	8
9	5	4	3	8	2	7	6	1
7	3	8	6	9	1	5	2	4

EASY - 16

1	4	9	3	5	8	7	6	2
7	2	6	1	9	4	5	8	3
5	3	8	6	2	7	1	4	9
9	1	3	5	8	6	4	2	7
6	7	4	2	1	3	8	9	5
8	5	2	7	4	9	3	1	6
3	9	7	4	6	1	2	5	8
4	8	5	9	7	2	6	3	1
2	6	1	8	3	5	9	7	4

EASY - 17

4	6	5	9	2	1	8	3	7
8	9	3	7	4	5	2	1	6
1	7	2	6	8	3	5	4	9
7	4	8	1	5	6	9	2	3
5	3	1	4	9	2	6	7	8
9	2	6	8	3	7	4	5	1
2	8	9	3	7	4	1	6	5
6	5	7	2	1	9	3	8	4
3	1	4	5	6	8	7	9	2

EASY - 18

6	3	2	5	7	1	9	8	
9	7	5	8	6	4	2	3	
1	4	8	9	2	3	5	7	
4	2	6	1	9	7	8	5	
8	5	3	2	4	6	7	1	
7	1	9	3	8	5	6	4	
5	9	7	4	3	2	1	6	
2	6	4	7	1	8	3	9	
3	8	1	6	5	9	4	2	

EASY - 19

1	2	3	9	4	7	6	8	5
7	5	8	6	1	2	4	3	9
4	6	9	8	5	3	1	2	7
3	8	6	5	7	9	2	1	4
5	7	1	2	6	4	8	9	3
2	9	4	3	8	1	5	7	6
6	3	5	7	2	8	9	4	1
9	4	2	1	3	5	7	6	8
8	1	7	4	9	6	3	5	2

EASY - 20

7	8	2	4	5	1	9	3	6
9	1	6	3	7	8	2	4	5
4	3	5	2	6	9	7	1	8
3	2	1	7	8	4	5	6	9
6	9	8	1	3	5	4	7	2
5	7	4	6	9	2	3	8	1
8	4	9	5	1	3	6	2	7
2	5	7	8	4	6	1	9	3
1	6	3	9	2	7	8	5	4

EASY - 21

4	6	7	2	3	8	1	9	5
9	1	8	5	7	6	4	3	2
2	5	3	1	4	9	7	8	6
7	3	9	8	2	1	6	5	4
6	8	4	7	5	3	9	2	1
5	2	1	6	9	4	3	7	8
8	9	5	4	1	7	2	6	3
3	4	6	9	8	2	5	1	7
1	7	2	3	6	5	8	4	9

EASY - 22

2	7	8	4	6	5	1	9	3
4	1	6	3	2	9	7	5	8
9	5	3	1	7	8	2	6	4
5	4	1	8	9	6	3	7	2
6	2	9	7	3	4	5	8	1
8	3	7	2	5	1	9	4	6
7	8	5	6	1	2	4	3	9
3	6	2	9	4	7	8	1	5
1	9	4	5	8	3	6	2	7

EASY - 23

2	1	7	6	3	4	9	8	5
5	6	9	1	8	7	4	2	3
4	3	8	9	5	2	7	6	1
1	5	3	7	6	9	8	4	2
7	8	6	2	4	5	1	3	9
9	4	2	8	1	3	5	7	6
3	9	5	4	7	6	2	1	8
6	7	1	5	2	8	3	9	4
8	2	4	3	9	1	6	5	7

EASY - 24

1	4	2	5	7	9	3	8	
7	6	5	8	3	1	2	4	
3	8	9	6	4	2	5	7	
8	9	3	4	2	7	6	1	
5	1	4	9	6	3	8	2	
6	2	7	1	8	5	9	3	
4	3	8	7	9	6	1	5	
9	7	1	2	5	8	4	6	
2	5	6	3	1	4	7	9	

EASY - 25

6	1	9	3	8	7	2	5	4
5	3	8	4	6	2	7	9	1
4	2	7	1	9	5	6	8	3
8	6	4	2	5	3	9	1	7
1	7	3	8	4	9	5	6	2
9	5	2	7	1	6	3	4	8
7	4	5	6	3	1	8	2	9
3	8	6	9	2	4	1	7	5
2	9	1	5	7	8	4	3	6

EASY - 26

2	5	4	9	1	8	6	7	3
1	9	7	3	2	6	8	4	5
3	8	6	7	4	5	2	9	1
5	2	9	8	6	4	1	3	7
8	4	3	1	9	7	5	2	6
7	6	1	2	5	3	9	8	4
9	7	5	6	3	2	4	1	8
4	1	8	5	7	9	3	6	2
6	3	2	4	8	1	7	5	9

EASY - 27

4	3	8	7	1	9	5	6	2
6	2	7	3	5	4	1	8	9
1	9	5	8	6	2	3	4	7
5	7	2	4	3	6	8	9	1
8	6	9	1	7	5	2	3	4
3	4	1	9	2	8	6	7	5
7	8	3	2	4	1	9	5	6
9	1	6	5	8	7	4	2	3
2	5	4	6	9	3	7	1	8

EASY - 28

3	9	5	1	4	2	7	6	8
7	8	2	6	9	3	1	4	5
4	1	6	5	7	8	3	2	9
6	4	9	3	2	7	8	5	1
1	2	8	9	5	4	6	3	7
5	3	7	8	1	6	4	9	2
2	7	1	4	6	9	5	8	3
8	5	4	2	3	1	9	7	6
9	6	3	7	8	5	2	1	4

EASY - 29

4	3	8	1	2	5	7	9	6
7	6	9	3	8	4	1	5	2
5	1	2	7	6	9	4	3	8
8	5	1	2	4	7	9	6	3
3	9	4	5	1	6	2	8	7
2	7	6	8	9	3	5	1	4
9	4	5	6	3	2	8	7	1
6	8	7	4	5	1	3	2	9
1	2	3	9	7	8	6	4	5

EASY - 30

8	6	7	1	3	9	4	5	
2	3	1	5	4	7	9	8	
5	9	4	8	2	6	7	3	
6	8	9	4	1	2	5	7	
3	7	2	6	8	5	1	4	
1	4	5	7	9	3	6	2	
7	2	3	9	6	4	8	1	
9	5	8	2	7	1	3	6	
4	1	6	3	5	8	2	9	

EASY - 31

9	1	7	5	6	2	3	4	8
6	3	4	7	9	8	1	2	5
5	2	8	3	1	4	7	9	6
8	4	3	9	2	1	5	6	7
7	9	1	6	4	5	2	8	3
2	6	5	8	3	7	9	1	4
3	8	6	2	5	9	4	7	1
4	7	9	1	8	3	6	5	2
1	5	2	4	7	6	8	3	9

EASY - 32

6	5	4	3	9	8	1	2	7
7	8	2	4	5	1	9	3	6
3	9	1	2	7	6	5	4	8
5	6	3	8	1	4	2	7	9
2	1	8	9	3	7	6	5	4
9	4	7	6	2	5	8	1	3
4	7	5	1	8	9	3	6	2
8	3	6	5	4	2	7	9	1
1	2	9	7	6	3	4	8	5

EASY - 33

3	4	8	1	9	2	5	7	6
1	7	6	4	8	5	3	9	2
9	2	5	6	7	3	8	1	4
7	9	3	5	1	6	2	4	8
8	5	2	9	3	4	1	6	7
4	6	1	8	2	7	9	3	5
5	1	4	3	6	8	7	2	9
2	8	9	7	4	1	6	5	3
6	3	7	2	5	9	4	8	1

EASY - 34

9	8	3	2	1	4	7	6	5
7	2	5	3	9	6	4	1	8
1	4	6	5	7	8	3	2	9
2	6	9	7	8	3	5	4	1
8	5	1	6	4	9	2	3	7
4	3	7	1	5	2	9	8	6
3	9	2	8	6	5	1	7	4
5	7	8	4	3	1	6	9	2
6	1	4	9	2	7	8	5	3

EASY - 35

9	1	8	7	4	3	2	6	5
2	6	7	9	8	5	1	4	3
3	5	4	1	2	6	7	9	8
1	9	2	5	3	8	6	7	4
4	3	5	6	9	7	8	2	1
8	7	6	4	1	2	3	5	9
7	4	9	8	6	1	5	3	2
5	8	3	2	7	9	4	1	6
6	2	1	3	5	4	9	8	7

EASY - 36

2	6	9	7	4	8	3	1	
8	3	4	1	5	9	6	7	
5	1	7	6	2	3	9	8	
7	2	8	4	9	1	5	3	
4	5	1	3	6	2	8	9	
6	9	3	5	8	7	2	4	
3	4	5	9	7	6	1	2	
1	7	2	8	3	5	4	6	
9	8	6	2	1	4	7	5	

EASY - 37

1	6	9	7	4	2	5	8	3
5	2	3	8	1	9	4	6	7
7	4	8	5	3	6	1	2	9
4	7	5	3	9	8	2	1	6
2	3	6	1	7	5	9	4	8
8	9	1	6	2	4	3	7	5
9	8	2	4	5	7	6	3	1
3	5	7	2	6	1	8	9	4
6	1	4	9	8	3	7	5	2

EASY - 38

5	8	6	9	3	2	4	7	1
1	4	2	6	5	7	3	8	9
9	7	3	1	4	8	2	6	5
3	2	1	8	7	6	9	5	4
6	5	7	3	9	4	8	1	2
8	9	4	5	2	1	7	3	6
7	1	5	4	8	9	6	2	3
2	3	9	7	6	5	1	4	8
4	6	8	2	1	3	5	9	7

EASY - 39

9	4	6	1	7	5	8	3	2
3	1	7	2	9	8	5	4	6
2	5	8	3	6	4	1	9	7
5	9	2	8	4	3	6	7	1
7	8	4	5	1	6	3	2	9
6	3	1	9	2	7	4	5	8
4	6	9	7	3	1	2	8	5
8	2	3	6	5	9	7	1	4
1	7	5	4	8	2	9	6	3

EASY - 40

7	1	6	4	8	5	3	2	9
4	5	8	3	2	9	6	7	1
9	2	3	1	6	7	4	5	8
2	8	1	7	5	6	9	3	4
3	4	9	8	1	2	7	6	5
5	6	7	9	3	4	8	1	2
1	7	4	5	9	3	2	8	6
6	9	5	2	7	8	1	4	3
8	3	2	6	4	1	5	9	7

EASY - 41

6	8	4	1	7	5	2	9	3
9	1	3	2	4	6	8	5	7
5	7	2	3	8	9	1	6	4
3	6	1	5	9	2	4	7	8
7	4	9	6	1	8	3	2	5
8	2	5	4	3	7	9	1	6
1	5	7	8	2	4	6	3	9
2	9	8	7	6	3	5	4	1
4	3	6	9	5	1	7	8	2

EASY - 42

2	6	4	3	1	8	9	5	
5	3	1	4	9	7	8	2	
7	9	8	6	5	2	3	4	
4	7	3	1	2	9	5	6	
8	2	5	7	6	4	1	3	
9	1	6	8	3	5	2	7	
3	4	9	2	7	1	6	8	
6	5	7	9	8	3	4	1	
1	8	2	5	4	6	7	9	

EASY - 43

1	9	4	7	3	2	5	6	8
3	6	7	5	4	8	1	2	9
5	2	8	9	6	1	4	7	3
4	3	2	6	8	9	7	1	5
9	8	1	4	5	7	6	3	2
6	7	5	1	2	3	9	8	4
2	4	6	3	1	5	8	9	7
8	1	9	2	7	4	3	5	6
7	5	3	8	9	6	2	4	1

EASY - 44

7	9	4	3	2	5	8	6	1
5	6	2	9	1	8	4	7	3
8	1	3	4	7	6	9	2	5
9	8	6	1	5	4	7	3	2
3	5	1	2	8	7	6	9	4
2	4	7	6	3	9	5	1	8
4	3	5	7	9	2	1	8	6
6	2	9	8	4	1	3	5	7
1	7	8	5	6	3	2	4	9

EASY - 45

4	8	9	7	1	3	6	5	2
6	1	3	2	5	4	9	8	7
7	5	2	8	6	9	1	4	3
3	7	5	9	2	6	4	1	8
2	9	8	4	7	1	5	3	6
1	4	6	5	3	8	7	2	9
5	3	7	6	4	2	8	9	1
9	2	4	1	8	7	3	6	5
8	6	1	3	9	5	2	7	4

EASY - 46

4	1	7	8	9	6	5	2	3
9	8	2	3	4	5	6	1	7
3	6	5	1	7	2	8	4	9
6	5	4	7	2	1	3	9	8
2	7	3	5	8	9	1	6	4
8	9	1	6	3	4	7	5	2
7	2	6	4	1	3	9	8	5
1	4	8	9	5	7	2	3	6
5	3	9	2	6	8	4	7	1

EASY - 47

4	9	7	6	2	8	5	3	1
5	8	2	3	4	1	7	9	6
3	6	1	7	9	5	8	4	2
2	3	9	8	5	7	1	6	4
6	4	8	1	3	2	9	7	5
1	7	5	4	6	9	3	2	8
9	5	3	2	1	4	6	8	7
7	2	6	5	8	3	4	1	9
8	1	4	9	7	6	2	5	3

EASY - 48

9	5	8	3	1	4	7	6	
6	3	4	7	9	2	5	8	
7	1	2	6	5	8	3	9	
5	6	7	2	4	9	8	1	
1	2	3	5	8	6	4	7	
8	4	9	1	3	7	6	2	
3	7	6	9	2	5	1	4	
4	9	1	8	6	3	2	5	
2	8	5	4	7	1	9	3	

EASY - 49

4	8	6	9	1	7	5	2	3
1	5	2	8	3	4	7	9	6
7	9	3	6	2	5	8	4	1
9	1	7	3	8	2	4	6	5
6	3	8	5	4	9	1	7	2
5	2	4	1	7	6	9	3	8
2	7	5	4	6	1	3	8	9
8	6	1	7	9	3	2	5	4
3	4	9	2	5	8	6	1	7

EASY - 50

6	8	5	2	4	7	9	3	1
1	9	2	8	3	6	5	4	7
4	7	3	9	5	1	6	2	8
7	2	4	3	9	8	1	6	5
3	5	6	1	7	4	2	8	9
9	1	8	6	2	5	3	7	4
2	4	7	5	1	3	8	9	6
5	6	9	7	8	2	4	1	3
8	3	1	4	6	9	7	5	2

EASY - 51

3	1	5	2	4	9	7	6	8
6	4	8	5	7	3	9	2	1
7	9	2	6	1	8	4	5	3
8	3	9	1	6	4	5	7	2
1	5	6	7	8	2	3	9	4
4	2	7	9	3	5	8	1	6
2	6	4	8	9	7	1	3	5
5	7	3	4	2	1	6	8	9
9	8	1	3	5	6	2	4	7

EASY - 52

4	6	1	9	7	3	8	2	5
3	8	2	1	4	5	7	9	6
7	5	9	2	6	8	1	4	3
1	9	7	5	8	4	6	3	2
6	2	4	7	3	9	5	8	1
8	3	5	6	2	1	4	7	9
5	1	3	8	9	7	2	6	4
2	4	8	3	5	6	9	1	7
9	7	6	4	1	2	3	5	8

EASY - 53

6	4	5	3	7	1	2	8	9
1	2	7	6	9	8	5	3	4
8	9	3	4	5	2	7	6	1
7	1	2	8	3	6	9	4	5
3	8	4	5	2	9	1	7	6
9	5	6	1	4	7	8	2	3
4	3	8	7	1	5	6	9	2
2	7	1	9	6	3	4	5	8
5	6	9	2	8	4	3	1	7

EASY - 54

1	5	8	4	9	3	6	2	7
4	7	6	5	8	2	9	1	3
9	3	2	6	1	7	5	4	8
8	2	9	3	6	4	1	7	5
6	1	5	2	7	8	3	9	4
3	4	7	1	5	9	2	8	6
2	9	3	7	4	6	8	5	1
7	8	1	9	3	5	4	6	2
5	6	4	8	2	1	7	3	9

EASY - 55

1	9	4	5	7	8	3	6	2
5	3	7	6	2	9	4	8	1
6	2	8	3	4	1	7	9	5
8	5	3	1	6	4	9	2	7
9	7	1	2	8	3	6	5	4
4	6	2	7	9	5	1	3	8
7	1	9	8	5	6	2	4	3
2	8	6	4	3	7	5	1	9
3	4	5	9	1	2	8	7	6

EASY - 56

8	1	9	4	2	7	6	3	5
7	6	4	1	5	3	2	8	9
5	3	2	6	8	9	7	1	4
4	5	6	2	7	1	8	9	3
2	8	1	3	9	6	5	4	7
9	7	3	8	4	5	1	2	6
1	9	8	7	6	4	3	5	2
6	2	5	9	3	8	4	7	1
3	4	7	5	1	2	9	6	8

EASY - 57

7	6	9	5	2	8	1	3	4
2	8	1	6	3	4	7	9	5
5	4	3	9	7	1	6	2	8
8	5	2	1	4	3	9	6	7
9	3	4	8	6	7	5	1	2
1	7	6	2	5	9	8	4	3
4	9	5	7	1	2	3	8	6
3	1	7	4	8	6	2	5	9
6	2	8	3	9	5	4	7	1

EASY - 58

7	1	5	2	6	9	8	4	3
2	6	4	8	1	3	7	9	5
8	9	3	7	4	5	2	6	1
1	2	7	5	9	8	6	3	4
5	4	6	3	7	1	9	2	8
9	3	8	4	2	6	5	1	7
6	8	2	1	3	7	4	5	9
3	5	9	6	8	4	1	7	2
4	7	1	9	5	2	3	8	6

EASY - 59

3	1	2	4	8	9	7	6	5
8	5	4	6	2	7	3	1	9
6	7	9	5	1	3	8	4	2
9	3	7	8	5	6	4	2	1
2	6	1	9	7	4	5	8	3
4	8	5	1	3	2	6	9	7
7	2	6	3	4	1	9	5	8
5	4	3	2	9	8	1	7	6
1	9	8	7	6	5	2	3	4

EASY - 60

4	7	3	9	2	5	8	1	6
6	1	8	3	4	7	9	5	2
2	5	9	6	1	8	7	4	3
3	8	4	5	6	9	1	2	7
5	2	7	1	3	4	6	8	9
9	6	1	8	7	2	5	3	4
8	4	5	7	9	3	2	6	1
7	3	6	2	8	1	4	9	5
1	9	2	4	5	6	3	7	8

EASY - 61

6	7	2	3	8	5	1	4	9
3	9	4	7	1	2	6	5	8
5	8	1	9	6	4	7	3	2
2	4	6	8	9	7	3	1	5
8	1	7	5	4	3	2	9	6
9	5	3	6	2	1	8	7	4
7	6	8	1	5	9	4	2	3
1	2	5	4	3	6	9	8	7
4	3	9	2	7	8	5	6	1

EASY - 62

4	7	5	2	1	8	3	9	6
2	3	9	6	4	7	8	5	1
1	6	8	9	3	5	2	4	7
7	5	3	8	2	4	6	1	9
8	9	1	7	5	6	4	3	2
6	4	2	1	9	3	5	7	8
5	1	7	3	8	2	9	6	4
3	8	6	4	7	9	1	2	5
9	2	4	5	6	1	7	8	3

EASY - 63

4	3	5	1	6	8	9	7	2
8	9	6	2	3	7	5	4	1
2	7	1	5	9	4	8	6	3
6	1	8	3	7	5	2	9	4
7	2	4	9	8	1	6	3	5
9	5	3	6	4	2	1	8	7
3	8	2	7	5	6	4	1	9
1	4	7	8	2	9	3	5	6
5	6	9	4	1	3	7	2	8

EASY - 64

3	4	6	1	2	8	5	9	7
2	1	9	7	5	4	6	8	3
8	7	5	3	6	9	4	1	2
9	5	4	2	7	6	1	3	8
7	3	8	9	4	1	2	5	6
6	2	1	8	3	5	9	7	4
4	6	7	5	1	3	8	2	9
1	9	2	6	8	7	3	4	5
5	8	3	4	9	2	7	6	1

EASY - 65

6	2	3	8	1	4	7	9	5
5	8	9	7	3	6	1	4	2
1	7	4	5	9	2	8	3	6
3	9	1	2	4	8	6	5	7
7	5	2	3	6	1	9	8	4
4	6	8	9	7	5	3	2	1
2	4	6	1	8	9	5	7	3
9	3	5	6	2	7	4	1	8
8	1	7	4	5	3	2	6	9

EASY - 66

9	2	8	3	1	4	6	7	5
6	1	5	2	7	8	3	9	4
4	3	7	5	9	6	2	1	8
8	5	1	7	6	3	9	4	2
3	6	9	4	5	2	1	8	7
7	4	2	9	8	1	5	6	3
1	7	6	8	3	5	4	2	9
5	9	4	1	2	7	8	3	6
2	8	3	6	4	9	7	5	1

EASY - 67

9	5	1	6	7	2	3	8	4
6	4	3	5	1	8	2	7	9
2	8	7	4	9	3	5	1	6
7	9	8	3	4	6	1	2	5
1	6	2	7	5	9	8	4	3
5	3	4	8	2	1	6	9	7
8	1	5	9	3	4	7	6	2
3	2	9	1	6	7	4	5	8
4	7	6	2	8	5	9	3	1

EASY - 68

8	1	2	7	4	6	3	5	9
7	9	5	3	8	2	6	1	4
3	6	4	1	5	9	7	8	2
2	7	3	8	9	1	5	4	6
9	5	8	4	6	7	2	3	1
6	4	1	2	3	5	8	9	7
5	2	9	6	1	8	4	7	3
1	3	6	5	7	4	9	2	8
4	8	7	9	2	3	1	6	5

EASY - 69

6	9	1	3	7	2	5	8	4
7	3	2	8	5	4	1	9	6
5	4	8	6	9	1	3	2	7
2	5	3	4	1	6	8	7	9
9	1	6	7	3	8	2	4	5
4	8	7	9	2	5	6	3	1
8	2	9	5	6	7	4	1	3
1	7	5	2	4	3	9	6	8
3	6	4	1	8	9	7	5	2

EASY - 70

7	4	8	5	2	9	3	1	6
2	5	9	6	1	3	8	7	4
1	6	3	7	4	8	2	9	5
5	2	7	3	8	1	4	6	9
9	8	6	4	5	7	1	2	3
4	3	1	2	9	6	5	8	7
8	7	5	9	3	2	6	4	1
3	9	2	1	6	4	7	5	8
6	1	4	8	7	5	9	3	2

EASY - 71

2	7	4	5	3	8	9	6	1
5	9	6	1	2	4	7	3	8
1	3	8	6	9	7	4	2	5
4	2	9	8	1	3	6	5	7
7	6	1	2	4	5	3	8	9
8	5	3	9	7	6	2	1	4
3	8	5	4	6	9	1	7	2
6	4	2	7	5	1	8	9	3
9	1	7	3	8	2	5	4	6

EASY - 72

3	5	6	2	8	7	4	1	9
4	9	2	1	6	3	8	5	7
8	1	7	9	5	4	3	2	6
1	2	4	6	9	8	7	3	5
5	7	3	4	1	2	9	6	8
9	6	8	7	3	5	2	4	1
6	4	5	8	2	9	1	7	3
2	8	1	3	7	6	5	9	4
7	3	9	5	4	1	6	8	2

EASY - 73

8	7	2	9	5	3	4	6	1
6	9	4	8	2	1	3	5	7
1	3	5	4	6	7	2	9	8
9	6	8	5	1	2	7	3	4
4	5	3	7	8	9	6	1	2
2	1	7	3	4	6	9	8	5
7	4	1	6	3	5	8	2	9
3	2	9	1	7	8	5	4	6
5	8	6	2	9	4	1	7	3

EASY - 74

7	9	4	2	3	6	1	5	8
2	1	5	9	4	8	7	3	6
8	6	3	1	5	7	2	9	4
1	4	8	7	2	5	9	6	3
9	5	2	6	1	3	4	8	7
6	3	7	8	9	4	5	1	2
3	7	1	5	6	2	8	4	9
4	8	9	3	7	1	6	2	5
5	2	6	4	8	9	3	7	1

EASY - 75

2	1	5	8	4	6	3	9	7
4	7	6	3	2	9	1	5	8
8	9	3	5	1	7	4	6	2
1	6	2	7	3	5	8	4	9
3	4	8	2	9	1	5	7	6
9	5	7	4	6	8	2	1	3
5	8	4	6	7	3	9	2	1
7	2	9	1	8	4	6	3	5
6	3	1	9	5	2	7	8	4

EASY - 76

4	5	6	2	1	3	7	8	9
2	8	7	6	9	4	1	3	5
1	9	3	7	8	5	4	2	6
7	1	4	8	5	9	2	6	3
3	6	8	1	2	7	5	9	4
5	2	9	4	3	6	8	7	1
6	7	1	9	4	2	3	5	8
8	3	2	5	6	1	9	4	7
9	4	5	3	7	8	6	1	2

EASY - 77

8	2	4	3	5	7	9	1	6
5	6	7	1	9	4	3	2	8
3	1	9	6	8	2	7	5	4
1	9	6	5	7	3	4	8	2
2	8	3	9	4	1	6	7	5
7	4	5	2	6	8	1	9	3
4	3	8	7	2	9	5	6	1
9	5	1	8	3	6	2	4	7
6	7	2	4	1	5	8	3	9

EASY - 78

1	3	8	9	6	2	4	5	7
7	9	2	5	4	1	3	6	8
6	4	5	3	8	7	2	9	1
3	2	6	1	5	9	7	8	4
8	1	9	7	2	4	6	3	5
4	5	7	6	3	8	1	2	9
9	8	3	4	1	6	5	7	2
5	7	4	2	9	3	8	1	6
2	6	1	8	7	5	9	4	3

EASY - 79

1	6	9	2	7	3	4	8	5
4	8	2	5	9	1	6	3	7
7	3	5	6	8	4	1	9	2
2	9	4	7	1	5	3	6	8
8	7	1	4	3	6	5	2	9
3	5	6	9	2	8	7	4	1
6	4	7	8	5	2	9	1	3
5	2	3	1	4	9	8	7	6
9	1	8	3	6	7	2	5	4

EASY - 80

2	3	4	6	1	7	5	9	8
7	9	8	3	5	4	1	6	2
1	5	6	2	9	8	7	4	3
8	4	5	1	7	6	3	2	9
3	7	1	4	2	9	8	5	6
9	6	2	5	8	3	4	1	7
4	1	9	7	3	2	6	8	5
5	8	7	9	6	1	2	3	4
6	2	3	8	4	5	9	7	1

EASY - 81

2	5	6	8	3	7	4	1	9
8	7	1	2	4	9	6	3	5
3	9	4	5	6	1	7	8	2
6	1	2	3	7	4	9	5	8
5	4	7	9	2	8	3	6	1
9	8	3	6	1	5	2	4	7
1	6	9	7	8	3	5	2	4
7	3	8	4	5	2	1	9	6
4	2	5	1	9	6	8	7	3

EASY - 82

2	4	7	3	9	1	6	5	8
3	9	6	8	2	5	1	7	4
8	5	1	7	4	6	3	2	9
4	1	8	9	5	7	2	6	3
9	6	5	2	3	4	8	1	7
7	2	3	1	6	8	4	9	5
5	8	9	4	1	2	7	3	6
1	3	4	6	7	9	5	8	2
6	7	2	5	8	3	9	4	1

EASY - 83

8	5	7	6	3	4	9	1	2
3	4	9	1	7	2	6	5	8
6	1	2	9	5	8	7	3	4
7	8	1	2	6	3	4	9	5
4	2	5	7	8	9	1	6	3
9	6	3	4	1	5	8	2	7
1	3	6	5	4	7	2	8	9
2	7	8	3	9	1	5	4	6
5	9	4	8	2	6	3	7	1

EASY - 84

5	7	8	6	2	9	4	3	1
1	6	4	5	3	7	8	9	2
3	2	9	8	4	1	5	7	6
4	1	7	9	6	8	2	5	3
6	9	5	2	7	3	1	4	8
2	8	3	4	1	5	9	6	7
8	4	6	7	9	2	3	1	5
9	5	1	3	8	6	7	2	4
7	3	2	1	5	4	6	8	9

EASY - 85

5	9	6	8	3	1	7	4	2
3	2	8	4	7	9	5	1	6
1	4	7	2	6	5	8	3	9
2	6	5	7	1	3	9	8	4
9	7	1	6	4	8	2	5	3
4	8	3	9	5	2	6	7	1
8	1	2	3	9	7	4	6	5
7	5	4	1	2	6	3	9	8
6	3	9	5	8	4	1	2	7

EASY - 86

4	6	5	2	8	9	3	7	1
1	2	9	3	4	7	8	6	5
3	7	8	5	1	6	4	2	9
8	5	4	9	6	1	2	3	7
6	9	2	8	7	3	5	1	4
7	1	3	4	2	5	9	8	6
5	8	7	1	9	2	6	4	3
9	4	1	6	3	8	7	5	2
2	3	6	7	5	4	1	9	8

EASY - 87

3	2	7	5	6	4	1	8	9
4	1	6	9	8	3	7	2	5
8	5	9	2	1	7	4	6	3
2	9	8	6	7	5	3	4	1
6	3	1	4	9	8	5	7	2
5	7	4	3	2	1	8	9	6
9	8	5	7	3	6	2	1	4
1	6	3	8	4	2	9	5	7
7	4	2	1	5	9	6	3	8

EASY - 88

9	7	8	3	1	6	2	5	4
3	4	5	9	7	2	8	6	1
1	6	2	5	4	8	3	9	7
5	8	6	1	2	9	7	4	3
2	3	7	4	8	5	9	1	6
4	1	9	6	3	7	5	2	8
8	5	3	2	6	1	4	7	9
6	9	4	7	5	3	1	8	2
7	2	1	8	9	4	6	3	5

EASY - 89

4	2	8	3	5	1	7	9	6
1	9	3	2	6	7	4	5	8
5	6	7	9	8	4	2	1	3
6	8	2	4	7	9	1	3	5
7	4	1	8	3	5	9	6	2
3	5	9	6	1	2	8	4	7
8	3	4	1	2	6	5	7	9
9	7	6	5	4	8	3	2	1
2	1	5	7	9	3	6	8	4

EASY - 90

2	1	8	9	3	4	6	5	7
3	4	7	5	6	2	9	1	8
9	5	6	1	7	8	2	4	3
8	2	9	6	4	5	7	3	1
7	3	4	8	2	1	5	9	6
5	6	1	3	9	7	4	8	2
1	8	2	7	5	9	3	6	4
4	9	3	2	1	6	8	7	5
6	7	5	4	8	3	1	2	9

EASY - 91

7	8	9	6	2	5	3	4	1
6	4	1	7	3	9	2	8	5
3	5	2	4	1	8	9	7	6
1	7	4	2	9	3	5	6	8
5	9	8	1	4	6	7	3	2
2	3	6	8	5	7	4	1	9
8	2	3	9	7	1	6	5	4
9	1	7	5	6	4	8	2	3
4	6	5	3	8	2	1	9	7

EASY - 92

1	7	4	3	5	6	9	2	8
3	5	9	1	2	8	7	4	6
8	6	2	4	9	7	5	1	3
4	8	5	2	1	3	6	9	7
2	1	6	8	7	9	4	3	5
9	3	7	5	6	4	2	8	1
6	9	8	7	4	1	3	5	2
7	2	3	9	8	5	1	6	4
5	4	1	6	3	2	8	7	9

EASY - 93

4	2	6	5	1	9	8	7	3
1	9	7	4	8	3	2	5	6
3	5	8	7	6	2	4	9	1
9	7	5	2	3	1	6	4	8
6	3	4	8	9	7	5	1	2
8	1	2	6	5	4	9	3	7
7	6	9	3	2	5	1	8	4
5	8	3	1	4	6	7	2	9
2	4	1	9	7	8	3	6	5

EASY - 94

4	1	5	7	6	3	2	8	9
3	8	2	1	5	9	4	7	6
9	7	6	4	8	2	3	1	5
7	3	8	9	2	4	5	6	1
5	6	1	8	3	7	9	2	4
2	4	9	5	1	6	7	3	8
6	2	4	3	9	1	8	5	7
1	5	7	2	4	8	6	9	3
8	9	3	6	7	5	1	4	2

EASY - 95

5	6	8	9	7	4	1	2	3
9	4	2	1	5	3	6	8	7
1	3	7	8	6	2	9	4	5
7	9	4	6	8	1	3	5	2
2	8	1	3	9	5	4	7	6
3	5	6	2	4	7	8	9	1
4	7	3	5	1	8	2	6	9
8	1	9	7	2	6	5	3	4
6	2	5	4	3	9	7	1	8

EASY - 96

1	8	3	7	2	4	6	5	9
7	5	4	6	9	8	3	2	1
9	2	6	1	5	3	8	4	7
6	3	1	9	7	5	2	8	4
5	9	7	8	4	2	1	3	6
2	4	8	3	6	1	9	7	5
8	1	5	4	3	9	7	6	2
4	6	9	2	8	7	5	1	3
3	7	2	5	1	6	4	9	8

EASY - 97

1	9	6	2	5	3	4	7	8
3	8	4	7	1	6	2	5	9
5	2	7	8	9	4	1	6	3
7	5	3	6	2	8	9	1	4
6	4	9	3	7	1	5	8	2
8	1	2	9	4	5	7	3	6
9	3	1	4	6	7	8	2	5
4	7	8	5	3	2	6	9	1
2	6	5	1	8	9	3	4	7

EASY - 98

9	1	2	6	4	3	5	8	7
4	8	6	7	5	9	2	1	3
3	5	7	1	8	2	4	6	9
1	3	8	5	6	4	9	7	2
5	6	9	2	1	7	3	4	8
7	2	4	9	3	8	6	5	1
6	9	1	3	7	5	8	2	4
2	4	5	8	9	1	7	3	6
8	7	3	4	2	6	1	9	5

EASY - 99

4	9	1	3	8	5	6	2	7
2	3	8	6	4	7	5	9	1
7	5	6	9	1	2	3	8	4
5	6	9	7	2	4	8	1	3
1	4	7	8	5	3	2	6	9
8	2	3	1	9	6	4	7	5
3	7	5	2	6	1	9	4	8
6	8	4	5	7	9	1	3	2
9	1	2	4	3	8	7	5	6

EASY - 100

8	6	9	5	3	2	7	4	1
1	2	4	9	7	6	3	8	5
7	3	5	8	4	1	2	9	6
4	9	6	1	2	8	5	3	7
2	5	7	3	9	4	6	1	8
3	1	8	7	6	5	9	2	4
9	7	1	4	5	3	8	6	2
5	4	2	6	8	9	1	7	3
6	8	3	2	1	7	4	5	9

EASY - 101

1	6	3	4	2	5	9	8	7
9	4	7	1	8	3	5	6	2
8	2	5	6	7	9	3	1	4
5	1	2	7	3	6	8	4	9
3	9	4	8	1	2	7	5	6
7	8	6	9	5	4	2	3	1
6	3	8	2	9	1	4	7	5
2	5	1	3	4	7	6	9	8
4	7	9	5	6	8	1	2	3

EASY - 102

3	1	5	7	4	8	2	6	9
7	4	2	9	6	1	5	8	3
6	9	8	3	2	5	4	7	1
1	5	3	4	8	9	6	2	7
8	2	7	1	5	6	3	9	4
9	6	4	2	3	7	1	5	8
5	3	9	8	1	2	7	4	6
2	8	1	6	7	4	9	3	5
4	7	6	5	9	3	8	1	2

EASY - 103

9	2	7	3	4	1	8	5	6
1	6	4	5	9	8	7	2	3
3	5	8	7	2	6	9	4	1
8	3	1	9	7	4	2	6	5
5	4	9	6	1	2	3	8	7
2	7	6	8	5	3	1	9	4
7	1	2	4	6	9	5	3	8
4	8	5	2	3	7	6	1	9
6	9	3	1	8	5	4	7	2

EASY - 104

7	6	8	3	2	4	9	1	5
3	9	5	6	7	1	2	8	4
4	2	1	5	9	8	3	7	6
2	1	7	9	4	3	5	6	8
8	4	3	2	5	6	7	9	1
6	5	9	1	8	7	4	2	3
5	8	4	7	1	2	6	3	9
1	3	2	4	6	9	8	5	7
9	7	6	8	3	5	1	4	2

EASY - 105

9	8	1	7	5	3	4	6	2
6	7	5	4	9	2	3	1	8
2	4	3	1	6	8	9	7	5
5	1	6	9	3	7	2	8	4
8	3	9	2	1	4	6	5	7
4	2	7	6	8	5	1	9	3
1	5	8	3	2	9	7	4	6
3	6	4	5	7	1	8	2	9
7	9	2	8	4	6	5	3	1

EASY - 106

2	8	9	6	4	7	5	1	3
3	4	7	1	8	5	6	2	9
5	1	6	3	9	2	8	4	7
7	9	2	5	6	4	1	3	8
8	5	1	7	2	3	9	6	4
6	3	4	8	1	9	7	5	2
9	7	5	4	3	6	2	8	1
1	6	3	2	7	8	4	9	5
4	2	8	9	5	1	3	7	6

EASY - 107

8	7	3	6	9	1	5	2	4
2	1	4	8	5	3	9	6	7
9	5	6	2	7	4	1	8	3
3	4	7	5	1	8	6	9	2
6	9	8	3	2	7	4	1	5
1	2	5	9	4	6	3	7	8
5	8	1	4	6	2	7	3	9
4	6	2	7	3	9	8	5	1
7	3	9	1	8	5	2	4	6

EASY - 108

6	5	9	7	4	1	8	3	2
2	1	7	8	3	6	5	9	4
3	8	4	2	5	9	1	7	6
8	9	5	1	2	3	6	4	7
7	3	2	5	6	4	9	8	1
4	6	1	9	8	7	2	5	3
5	4	3	6	1	8	7	2	9
9	2	6	3	7	5	4	1	8
1	7	8	4	9	2	3	6	5

EASY - 109

7	8	9	1	5	3	4	6	2
3	1	6	2	7	4	5	9	8
2	4	5	9	6	8	7	3	1
5	9	1	7	3	2	8	4	6
4	2	3	6	8	5	1	7	9
6	7	8	4	9	1	2	5	3
1	3	2	5	4	6	9	8	7
9	6	4	8	1	7	3	2	5
8	5	7	3	2	9	6	1	4

EASY - 110

2	3	9	1	8	6	5	4	7
8	5	1	4	9	7	2	3	6
4	7	6	3	2	5	1	9	8
9	2	4	5	6	8	3	7	1
5	6	3	9	7	1	4	8	2
1	8	7	2	3	4	6	5	9
3	9	8	6	4	2	7	1	5
7	1	2	8	5	3	9	6	4
6	4	5	7	1	9	8	2	3

EASY - 111

8	5	1	6	7	2	3	9	4
9	6	3	4	8	5	1	2	7
4	2	7	3	1	9	8	6	5
2	3	5	1	9	7	6	4	8
7	1	4	8	5	6	9	3	2
6	9	8	2	3	4	7	5	1
1	8	9	5	2	3	4	7	6
3	4	2	7	6	1	5	8	9
5	7	6	9	4	8	2	1	3

EASY - 112

9	7	3	4	2	8	5	1	6
1	6	5	7	3	9	4	2	8
8	4	2	1	6	5	7	3	9
6	8	4	3	5	2	1	9	7
2	3	1	8	9	7	6	5	4
7	5	9	6	1	4	3	8	2
3	2	8	5	7	6	9	4	1
5	9	7	2	4	1	8	6	3
4	1	6	9	8	3	2	7	5

EASY - 113

1	2	7	4	6	3	8	9	5
3	5	4	9	1	8	2	6	7
8	6	9	7	2	5	1	4	3
2	1	5	3	9	6	4	7	8
7	8	3	5	4	2	6	1	9
9	4	6	1	8	7	3	5	2
4	3	1	8	7	9	5	2	6
6	9	8	2	5	4	7	3	1
5	7	2	6	3	1	9	8	4

EASY - 114

5	8	1	4	3	9	2	7	6
6	3	2	7	8	5	1	9	4
9	4	7	2	6	1	8	3	5
3	7	9	6	1	8	5	4	2
1	6	4	5	2	3	7	8	9
8	2	5	9	4	7	3	6	1
7	1	6	8	9	2	4	5	3
4	5	3	1	7	6	9	2	8
2	9	8	3	5	4	6	1	7

EASY - 115

2	8	5	7	4	1	9	3	6
4	1	9	5	3	6	7	8	2
3	7	6	2	9	8	5	1	4
6	2	4	9	1	3	8	7	5
8	5	1	6	7	2	3	4	9
7	9	3	8	5	4	6	2	1
1	6	7	4	8	5	2	9	3
9	4	2	3	6	7	1	5	8
5	3	8	1	2	9	4	6	7

EASY - 116

6	9	1	5	2	3	8	4	7
4	3	5	6	7	8	1	2	9
7	2	8	1	4	9	5	3	6
8	1	7	9	6	4	2	5	3
2	5	6	3	1	7	9	8	4
3	4	9	2	8	5	6	7	1
1	8	2	7	3	6	4	9	5
5	7	4	8	9	1	3	6	2
9	6	3	4	5	2	7	1	8

EASY - 117

5	8	2	6	9	7	4	1	3
7	1	4	3	2	8	6	5	9
6	3	9	4	5	1	2	8	7
4	5	8	1	7	6	9	3	2
9	7	6	2	3	5	8	4	1
3	2	1	8	4	9	5	7	6
8	4	7	9	6	3	1	2	5
2	9	5	7	1	4	3	6	8
1	6	3	5	8	2	7	9	4

EASY - 118

8	9	6	5	3	4	1	7	2
7	2	3	6	1	8	5	4	9
1	5	4	7	2	9	8	6	3
3	7	5	2	4	6	9	1	8
2	6	9	1	8	7	4	3	5
4	8	1	3	9	5	6	2	7
9	1	2	8	6	3	7	5	4
6	4	7	9	5	2	3	8	1
5	3	8	4	7	1	2	9	6

EASY - 119

5	8	1	7	4	3	9	6	2
7	2	6	9	8	1	5	4	3
9	3	4	6	5	2	7	8	1
6	7	5	4	1	9	3	2	8
8	9	3	5	2	6	4	1	7
4	1	2	3	7	8	6	5	9
3	6	8	2	9	4	1	7	5
2	5	9	1	6	7	8	3	4
1	4	7	8	3	5	2	9	6

EASY - 120

4	7	6	3	5	8	2	1	9
8	3	1	4	9	2	5	6	7
9	5	2	6	7	1	4	8	3
5	1	3	2	4	9	6	7	8
7	2	9	8	6	3	1	4	5
6	8	4	7	1	5	3	9	2
2	4	5	9	8	6	7	3	1
1	9	7	5	3	4	8	2	6
3	6	8	1	2	7	9	5	4

EASY - 121

4	2	1	5	7	8	3	6	9
5	6	3	1	9	2	4	7	8
8	9	7	3	6	4	2	5	1
3	8	2	7	5	6	9	1	4
7	5	9	4	3	1	6	8	2
1	4	6	8	2	9	5	3	7
6	7	4	9	1	5	8	2	3
2	3	8	6	4	7	1	9	5
9	1	5	2	8	3	7	4	6

EASY - 122

6	9	2	4	5	7	8	1	3
1	7	8	2	3	6	9	5	4
3	5	4	8	1	9	2	7	6
5	2	3	6	7	8	4	9	1
9	8	6	3	4	1	7	2	5
4	1	7	5	9	2	6	3	8
7	6	1	9	8	5	3	4	2
8	4	9	1	2	3	5	6	7
2	3	5	7	6	4	1	8	9

EASY - 123

2	5	4	6	3	1	8	7	9
6	1	3	7	9	8	2	4	5
8	9	7	5	4	2	1	3	6
5	8	1	3	7	4	9	6	2
4	6	9	2	1	5	7	8	3
3	7	2	9	8	6	4	5	1
1	2	5	4	6	7	3	9	8
9	4	6	8	2	3	5	1	7
7	3	8	1	5	9	6	2	4

EASY - 124

9	7	5	8	6	4	1	2	3
3	4	2	9	7	1	5	8	6
8	6	1	2	3	5	7	4	9
5	3	7	1	2	9	8	6	4
2	1	4	3	8	6	9	7	5
6	8	9	5	4	7	2	3	1
1	2	6	4	5	8	3	9	7
4	9	8	7	1	3	6	5	2
7	5	3	6	9	2	4	1	8

EASY - 125

4	2	1	3	9	5	8	7	6
5	3	9	6	7	8	4	2	1
6	8	7	2	4	1	9	3	5
8	5	2	4	6	7	1	9	3
9	4	3	1	5	2	6	8	7
7	1	6	9	8	3	2	5	4
2	6	8	5	3	4	7	1	9
3	7	4	8	1	9	5	6	2
1	9	5	7	2	6	3	4	8

EASY - 126

7	4	8	2	3	5	9	6	1
3	9	5	7	6	1	4	2	8
6	1	2	9	8	4	7	3	5
1	5	4	3	2	9	6	8	7
9	8	3	5	7	6	1	4	2
2	6	7	1	4	8	5	9	3
5	3	1	6	9	2	8	7	4
8	7	9	4	1	3	2	5	6
4	2	6	8	5	7	3	1	9

EASY - 127

1	7	4	9	5	8	6	3	2
9	6	2	3	7	4	8	5	1
8	5	3	1	2	6	9	4	7
2	4	8	6	1	3	7	9	5
5	3	9	7	8	2	1	6	4
6	1	7	4	9	5	2	8	3
3	9	5	2	6	7	4	1	8
4	2	6	8	3	1	5	7	9
7	8	1	5	4	9	3	2	6

EASY - 128

9	7	4	2	6	3	1	8	5
1	6	2	5	8	9	3	4	7
3	8	5	1	4	7	9	2	6
7	2	6	4	1	5	8	3	9
5	9	8	3	7	2	6	1	4
4	1	3	8	9	6	7	5	2
8	5	7	9	3	4	2	6	1
6	4	1	7	2	8	5	9	3
2	3	9	6	5	1	4	7	8

EASY - 129

3	2	9	5	1	4	8	6	7
5	7	8	3	6	9	1	2	4
6	1	4	8	2	7	5	3	9
2	8	6	7	4	3	9	5	1
1	9	7	6	5	8	3	4	2
4	3	5	2	9	1	7	8	6
8	6	2	1	7	5	4	9	3
9	5	1	4	3	2	6	7	8
7	4	3	9	8	6	2	1	5

EASY - 130

8	5	6	4	7	2	3	1	9
3	9	1	8	6	5	4	7	2
4	2	7	3	9	1	6	8	5
1	3	2	7	8	9	5	4	6
5	4	8	1	3	6	9	2	7
7	6	9	2	5	4	1	3	8
6	1	4	5	2	8	7	9	3
2	7	5	9	1	3	8	6	4
9	8	3	6	4	7	2	5	1

EASY - 131

4	3	2	9	8	7	5	6	1
7	8	6	1	5	2	3	4	9
9	1	5	3	6	4	2	8	7
8	7	9	6	2	3	4	1	5
1	2	4	8	7	5	9	3	6
6	5	3	4	9	1	7	2	8
2	6	8	5	4	9	1	7	3
3	9	7	2	1	8	6	5	4
5	4	1	7	3	6	8	9	2

EASY - 132

5	3	9	2	1	7	4	8	6
6	8	4	9	5	3	7	1	2
1	2	7	6	8	4	9	5	3
7	4	8	3	6	1	5	2	9
9	6	5	4	2	8	1	3	7
3	1	2	7	9	5	6	4	8
4	5	3	8	7	6	2	9	1
8	9	6	1	4	2	3	7	5
2	7	1	5	3	9	8	6	4

EASY - 133

9	2	8	4	7	1	6	5	3
7	4	6	5	3	2	8	1	9
5	1	3	8	9	6	4	7	2
6	3	5	7	8	9	2	4	1
4	8	1	6	2	5	9	3	7
2	7	9	1	4	3	5	6	8
8	6	4	9	1	7	3	2	5
3	9	7	2	5	4	1	8	6
1	5	2	3	6	8	7	9	4

EASY - 134

4	5	6	8	9	7	2	3	1
7	9	3	1	5	2	8	4	6
2	1	8	6	3	4	9	7	5
9	6	5	3	8	1	7	2	4
8	7	4	2	6	9	1	5	3
3	2	1	7	4	5	6	9	8
6	4	2	9	1	3	5	8	7
5	8	9	4	7	6	3	1	2
1	3	7	5	2	8	4	6	9

EASY - 135

6	1	3	7	5	8	2	4	9
4	7	5	2	9	6	3	1	8
9	8	2	4	3	1	5	6	7
1	4	7	5	6	3	8	9	2
2	5	8	1	7	9	4	3	6
3	6	9	8	2	4	7	5	1
5	3	6	9	8	2	1	7	4
7	2	1	6	4	5	9	8	3
8	9	4	3	1	7	6	2	5

EASY - 136

1	9	2	4	6	3	8	7	5
3	8	5	2	7	9	4	1	6
7	4	6	1	8	5	2	3	9
8	2	3	6	5	4	7	9	1
5	1	4	3	9	7	6	2	8
9	6	7	8	1	2	3	5	4
4	5	9	7	3	6	1	8	2
6	3	8	5	2	1	9	4	7
2	7	1	9	4	8	5	6	3

EASY - 137

1	5	6	2	7	8	3	4	9
2	3	9	1	5	4	6	7	8
7	8	4	3	9	6	5	2	1
6	2	3	4	8	7	1	9	5
5	1	7	9	6	2	4	8	3
4	9	8	5	1	3	2	6	7
3	4	1	7	2	9	8	5	6
9	6	5	8	4	1	7	3	2
8	7	2	6	3	5	9	1	4

EASY - 138

2	1	8	9	6	3	5	4	7
3	9	7	5	1	4	6	8	2
6	4	5	8	2	7	1	3	9
4	6	9	7	3	8	2	1	5
8	2	1	6	4	5	7	9	3
5	7	3	1	9	2	4	6	8
7	8	6	3	5	1	9	2	4
9	3	4	2	7	6	8	5	1
1	5	2	4	8	9	3	7	6

EASY - 139

6	3	7	9	1	4	5	8	2
1	8	4	2	5	3	7	6	9
5	9	2	6	7	8	4	3	1
4	7	9	1	8	2	3	5	6
2	5	3	7	4	6	9	1	8
8	6	1	5	3	9	2	4	7
3	2	6	8	9	5	1	7	4
9	1	5	4	6	7	8	2	3
7	4	8	3	2	1	6	9	5

EASY - 140

8	1	5	4	6	7	2	3	9
6	3	2	1	9	5	7	8	4
9	4	7	8	3	2	6	1	5
4	2	9	7	8	3	1	5	6
1	7	6	5	4	9	8	2	3
3	5	8	6	2	1	4	9	7
2	8	3	9	7	4	5	6	1
7	9	1	2	5	6	3	4	8
5	6	4	3	1	8	9	7	2

EASY - 141

6	9	8	7	5	2	3	1	4
4	5	2	3	6	1	7	8	9
3	1	7	4	9	8	6	5	2
2	3	1	5	4	9	8	6	7
7	4	9	6	8	3	1	2	5
8	6	5	1	2	7	9	4	3
9	8	6	2	3	5	4	7	1
1	2	3	8	7	4	5	9	6
5	7	4	9	1	6	2	3	8

EASY - 142

8	7	1	9	5	4	3	2	6
2	6	4	7	8	3	1	5	9
5	9	3	1	2	6	4	8	7
6	3	7	8	4	1	5	9	2
4	1	8	2	9	5	6	7	3
9	2	5	3	6	7	8	1	4
1	8	2	6	3	9	7	4	5
3	5	9	4	7	8	2	6	1
7	4	6	5	1	2	9	3	8

EASY - 143

8	4	6	1	2	5	9	7	3
9	5	7	6	4	3	8	2	1
3	2	1	8	7	9	6	4	5
7	8	2	9	5	6	1	3	4
5	3	9	7	1	4	2	6	8
1	6	4	2	3	8	5	9	7
6	7	8	4	9	1	3	5	2
4	9	5	3	8	2	7	1	6
2	1	3	5	6	7	4	8	9

EASY - 144

8	2	7	6	1	3	4	9	5
5	9	3	7	8	4	6	2	1
4	6	1	5	2	9	3	8	7
3	7	9	8	5	6	2	1	4
2	8	6	9	4	1	5	7	3
1	4	5	2	3	7	9	6	8
7	1	2	4	9	5	8	3	6
6	5	8	3	7	2	1	4	9
9	3	4	1	6	8	7	5	2

EASY - 145

5	2	9	7	3	8	1	6	4
1	7	6	9	5	4	3	2	8
3	8	4	2	1	6	5	9	7
6	5	1	8	7	9	4	3	2
8	4	3	5	2	1	9	7	6
7	9	2	6	4	3	8	5	1
2	1	8	3	6	5	7	4	9
9	6	5	4	8	7	2	1	3
4	3	7	1	9	2	6	8	5

EASY - 146

3	9	8	5	7	1	4	6	2
6	7	2	4	9	8	5	1	3
1	4	5	2	3	6	9	7	8
2	8	1	3	6	4	7	9	5
9	6	4	7	5	2	3	8	1
7	5	3	8	1	9	6	2	4
4	2	7	9	8	5	1	3	6
5	1	9	6	2	3	8	4	7
8	3	6	1	4	7	2	5	9

EASY - 147

7	4	5	6	8	2	9	3	1
1	8	9	4	3	5	7	2	6
2	3	6	9	7	1	8	4	5
6	7	1	2	5	4	3	9	8
5	9	4	8	6	3	1	7	2
8	2	3	7	1	9	6	5	4
9	5	7	1	2	8	4	6	3
3	6	8	5	4	7	2	1	9
4	1	2	3	9	6	5	8	7

EASY - 148

7	5	6	2	3	9	4	8	1
8	4	9	7	6	1	2	3	5
1	2	3	8	5	4	7	6	9
9	6	5	3	2	8	1	4	7
4	1	8	9	7	6	5	2	3
2	3	7	1	4	5	6	9	8
5	9	2	6	1	3	8	7	4
3	7	1	4	8	2	9	5	6
6	8	4	5	9	7	3	1	2

EASY - 149

4	8	5	7	2	9	1	6	3
6	9	3	1	5	8	4	7	2
1	2	7	4	6	3	9	8	5
9	7	6	2	1	4	5	3	8
2	3	8	6	9	5	7	1	4
5	4	1	3	8	7	2	9	6
3	6	2	5	7	1	8	4	9
7	5	9	8	4	6	3	2	1
8	1	4	9	3	2	6	5	7

EASY - 150

2	9	7	8	1	4	6	5	3
1	8	5	6	3	2	7	4	9
3	6	4	5	7	9	2	1	8
7	5	2	3	4	1	8	9	6
4	1	8	9	6	7	5	3	2
6	3	9	2	8	5	1	7	4
9	2	1	4	5	8	3	6	7
8	7	6	1	9	3	4	2	5
5	4	3	7	2	6	9	8	1

EASY - 151

8	5	3	4	2	1	7	6	9
6	1	9	5	7	8	4	3	2
4	2	7	3	9	6	1	5	8
3	6	8	7	1	5	2	9	4
2	9	1	8	6	4	3	7	5
5	7	4	9	3	2	6	8	1
1	3	5	6	4	9	8	2	7
7	8	2	1	5	3	9	4	6
9	4	6	2	8	7	5	1	3

EASY - 152

5	3	6	4	9	7	8	1	2
7	1	2	5	3	8	4	6	9
9	4	8	6	1	2	7	3	5
4	8	9	2	5	1	6	7	3
1	7	5	3	8	6	9	2	4
6	2	3	9	7	4	5	8	1
8	9	4	7	2	3	1	5	6
2	6	7	1	4	5	3	9	8
3	5	1	8	6	9	2	4	7

EASY - 153

8	2	9	5	6	3	7	1	4
3	6	4	1	7	2	9	8	5
5	7	1	9	4	8	6	3	2
7	4	5	6	3	1	2	9	8
2	9	8	4	5	7	1	6	3
6	1	3	2	8	9	5	4	7
1	8	2	3	9	5	4	7	6
9	3	6	7	2	4	8	5	1
4	5	7	8	1	6	3	2	9

EASY - 154

1	6	3	7	4	9	2	8	5
8	7	4	5	2	1	3	6	9
2	9	5	6	3	8	4	1	7
9	1	2	3	5	4	6	7	8
7	4	6	9	8	2	5	3	1
3	5	8	1	7	6	9	4	2
5	2	1	4	6	7	8	9	3
6	3	7	8	9	5	1	2	4
4	8	9	2	1	3	7	5	6

EASY - 155

1	2	9	5	7	6	3	8	4
6	7	8	9	3	4	1	2	5
5	3	4	2	1	8	6	7	9
9	6	5	4	2	3	7	1	8
3	4	1	7	8	5	9	6	2
7	8	2	1	6	9	5	4	3
4	1	7	3	5	2	8	9	6
2	5	6	8	9	1	4	3	7
8	9	3	6	4	7	2	5	1

EASY - 156

4	1	7	8	6	2	9	5	3
8	2	3	9	5	1	6	7	4
5	6	9	4	7	3	1	2	8
9	8	6	5	3	4	7	1	2
1	5	4	2	9	7	8	3	6
7	3	2	6	1	8	5	4	9
6	9	1	3	4	5	2	8	7
3	7	8	1	2	6	4	9	5
2	4	5	7	8	9	3	6	1

EASY - 157

7	4	2	3	8	9	6	1	5
8	1	3	6	5	2	7	4	9
9	6	5	7	4	1	8	3	2
1	9	6	8	2	7	4	5	3
5	8	4	1	3	6	9	2	7
3	2	7	5	9	4	1	6	8
6	5	1	2	7	8	3	9	4
4	3	8	9	1	5	2	7	6
2	7	9	4	6	3	5	8	1

EASY - 158

9	7	6	1	2	5	3	4	8
5	4	8	9	7	3	6	2	1
2	3	1	8	6	4	9	7	5
3	2	9	6	8	1	7	5	4
7	8	5	4	9	2	1	6	3
1	6	4	5	3	7	2	8	9
6	9	3	2	4	8	5	1	7
8	1	2	7	5	9	4	3	6
4	5	7	3	1	6	8	9	2

EASY - 159

4	5	6	7	1	3	9	2	8
2	7	3	9	8	4	6	5	1
9	8	1	5	2	6	7	3	4
1	6	9	3	4	2	8	7	5
7	3	5	8	9	1	4	6	2
8	4	2	6	7	5	3	1	9
3	9	8	1	5	7	2	4	6
6	1	4	2	3	8	5	9	7
5	2	7	4	6	9	1	8	3

EASY - 160

7	6	8	2	1	5	3	4	9
1	4	9	3	7	8	6	2	5
5	2	3	4	9	6	8	1	7
4	3	2	5	8	1	7	9	6
8	7	1	6	2	9	4	5	3
9	5	6	7	4	3	1	8	2
2	8	4	9	3	7	5	6	1
3	9	5	1	6	4	2	7	8
6	1	7	8	5	2	9	3	4

EASY - 161

7	8	3	5	4	2	6	9	1
2	1	4	7	6	9	8	3	5
5	6	9	8	3	1	4	7	2
4	9	1	3	2	6	5	8	7
3	2	7	1	5	8	9	6	4
8	5	6	9	7	4	1	2	3
9	7	5	6	1	3	2	4	8
6	3	2	4	8	5	7	1	9
1	4	8	2	9	7	3	5	6

EASY - 162

4	1	8	7	3	6	5	2	9
5	2	7	8	9	1	4	6	3
6	9	3	2	4	5	7	1	8
3	4	1	6	5	7	8	9	2
2	7	6	1	8	9	3	4	5
9	8	5	4	2	3	6	7	1
8	5	2	9	6	4	1	3	7
7	3	4	5	1	2	9	8	6
1	6	9	3	7	8	2	5	4

EASY - 163

9	7	4	5	2	8	6	3	1
1	3	5	6	9	4	2	7	8
2	8	6	1	3	7	5	9	4
7	2	1	9	8	3	4	5	6
4	6	8	7	5	2	3	1	9
5	9	3	4	1	6	8	2	7
3	4	2	8	7	9	1	6	5
6	5	7	3	4	1	9	8	2
8	1	9	2	6	5	7	4	3

EASY - 164

4	1	9	3	8	6	5	2	7
7	2	6	1	9	5	4	3	8
3	8	5	2	4	7	6	1	9
2	9	7	4	6	1	3	8	5
8	5	1	9	7	3	2	6	4
6	4	3	8	5	2	7	9	1
5	7	8	6	2	9	1	4	3
1	6	4	5	3	8	9	7	2
9	3	2	7	1	4	8	5	6

EASY - 165

4	7	8	6	9	5	3	2	1
3	1	9	7	8	2	4	6	5
2	6	5	3	1	4	7	8	9
8	2	6	9	4	3	1	5	7
1	9	7	8	5	6	2	4	3
5	3	4	1	2	7	6	9	8
9	4	1	2	3	8	5	7	6
7	8	2	5	6	1	9	3	4
6	5	3	4	7	9	8	1	2

EASY - 166

9	8	3	5	2	1	4	6	7
1	5	2	4	7	6	8	9	3
4	7	6	9	8	3	2	1	5
6	1	5	8	3	9	7	2	4
2	4	8	7	6	5	1	3	9
3	9	7	1	4	2	5	8	6
7	3	4	6	1	8	9	5	2
5	6	1	2	9	4	3	7	8
8	2	9	3	5	7	6	4	1

EASY - 167

8	4	5	3	6	9	1	7	2
1	3	2	7	8	4	6	9	5
7	9	6	1	2	5	8	3	4
9	6	4	8	1	2	7	5	3
2	5	7	6	4	3	9	1	8
3	1	8	9	5	7	2	4	6
5	8	3	2	9	1	4	6	7
4	2	1	5	7	6	3	8	9
6	7	9	4	3	8	5	2	1

EASY - 168

7	3	2	8	9	6	4	5	1
4	6	5	1	7	2	8	9	3
8	1	9	3	4	5	7	2	6
3	2	4	5	1	7	6	8	9
6	9	8	2	3	4	5	1	7
5	7	1	9	6	8	3	4	2
2	5	7	6	8	1	9	3	4
9	8	6	4	2	3	1	7	5
1	4	3	7	5	9	2	6	8

EASY - 169

9	5	8	7	2	4	6	1	3
4	1	2	3	8	6	9	7	5
6	7	3	1	9	5	8	4	2
8	3	7	2	4	9	1	5	6
1	9	6	8	5	3	4	2	7
5	2	4	6	1	7	3	9	8
2	4	9	5	6	8	7	3	1
3	8	5	9	7	1	2	6	4
7	6	1	4	3	2	5	8	9

EASY - 170

8	4	5	1	7	9	2	3	6
2	1	9	3	8	6	5	7	4
6	3	7	5	4	2	9	1	8
9	7	8	6	5	4	3	2	1
1	5	4	2	3	7	8	6	9
3	6	2	9	1	8	7	4	5
4	8	3	7	6	5	1	9	2
5	9	1	4	2	3	6	8	7
7	2	6	8	9	1	4	5	3

EASY - 171

4	7	8	5	6	9	1	3	2
3	5	9	7	1	2	8	4	6
2	6	1	3	8	4	9	5	7
6	8	5	2	4	1	7	9	3
9	3	4	6	7	5	2	1	8
7	1	2	8	9	3	4	6	5
1	2	6	4	3	7	5	8	9
5	4	3	9	2	8	6	7	1
8	9	7	1	5	6	3	2	4

EASY - 172

7	1	2	3	9	5	6	4	8
3	5	8	7	4	6	2	9	1
4	6	9	2	1	8	5	3	7
9	7	4	1	5	3	8	6	2
8	3	1	6	7	2	4	5	9
6	2	5	4	8	9	1	7	3
1	4	3	8	6	7	9	2	5
2	9	6	5	3	1	7	8	4
5	8	7	9	2	4	3	1	6

EASY - 173

2	6	3	7	1	8	4	9	5
9	5	4	2	3	6	7	8	1
7	8	1	4	5	9	3	2	6
6	2	9	1	4	5	8	7	3
3	7	5	8	9	2	1	6	4
4	1	8	6	7	3	2	5	9
1	3	6	9	2	7	5	4	8
5	9	7	3	8	4	6	1	2
8	4	2	5	6	1	9	3	7

EASY - 174

4	2	1	9	8	5	7	3	6
7	6	8	3	2	1	5	4	9
9	3	5	6	4	7	8	2	1
6	8	7	2	9	3	1	5	4
1	4	2	5	6	8	3	9	7
5	9	3	7	1	4	2	6	8
2	1	9	8	3	6	4	7	5
8	7	6	4	5	2	9	1	3
3	5	4	1	7	9	6	8	2

EASY - 175

7	3	4	9	1	5	8	2	6
9	6	1	7	2	8	5	3	4
8	5	2	3	6	4	7	1	9
4	2	8	5	9	1	6	7	3
5	7	3	6	4	2	9	8	1
6	1	9	8	7	3	4	5	2
1	8	5	4	3	9	2	6	7
3	4	7	2	8	6	1	9	5
2	9	6	1	5	7	3	4	8

EASY - 176

5	4	6	8	1	7	9	2	3
7	1	8	9	3	2	5	4	6
9	3	2	6	5	4	8	7	1
6	9	4	1	2	5	7	3	8
8	2	3	7	6	9	1	5	4
1	7	5	3	4	8	2	6	9
2	6	7	4	8	1	3	9	5
3	8	9	5	7	6	4	1	2
4	5	1	2	9	3	6	8	7

EASY - 177

4	6	1	2	9	5	3	7	8
3	8	9	7	1	4	2	5	6
2	5	7	8	3	6	4	9	1
1	7	4	5	6	2	8	3	9
5	9	6	3	4	8	7	1	2
8	3	2	1	7	9	5	6	4
6	1	8	4	5	7	9	2	3
7	2	3	9	8	1	6	4	5
9	4	5	6	2	3	1	8	7

EASY - 178

3	9	4	7	6	1	8	5	2
2	8	1	5	4	9	6	7	3
6	5	7	3	8	2	1	4	9
7	2	8	6	9	3	4	1	5
4	6	3	2	1	5	9	8	7
9	1	5	8	7	4	2	3	6
5	7	6	4	2	8	3	9	1
8	3	9	1	5	6	7	2	4
1	4	2	9	3	7	5	6	8

EASY - 179

9	7	2	6	5	3	4	8	1
8	6	1	2	9	4	7	5	3
4	3	5	8	1	7	2	6	9
6	8	4	9	7	1	3	2	5
2	9	3	4	8	5	6	1	7
5	1	7	3	6	2	8	9	4
3	2	6	5	4	9	1	7	8
7	4	9	1	2	8	5	3	6
1	5	8	7	3	6	9	4	2

EASY - 180

9	1	7	2	4	6	8	5	3
6	4	3	8	9	5	7	2	1
8	2	5	3	1	7	6	9	4
4	5	9	1	7	8	2	3	6
3	6	1	9	5	2	4	7	8
7	8	2	4	6	3	9	1	5
1	3	6	7	8	9	5	4	2
5	9	4	6	2	1	3	8	7
2	7	8	5	3	4	1	6	9

EASY - 181

8	9	6	3	7	1	2	4	5
4	1	3	9	5	2	8	6	7
7	2	5	4	8	6	9	3	1
1	8	7	5	4	9	6	2	3
5	6	4	2	1	3	7	9	8
9	3	2	8	6	7	5	1	4
2	4	9	7	3	5	1	8	6
3	5	1	6	9	8	4	7	2
6	7	8	1	2	4	3	5	9

EASY - 182

2	8	6	1	5	3	9	4	7
3	9	1	4	6	7	2	5	8
7	4	5	8	9	2	6	1	3
1	5	7	2	3	9	4	8	6
4	2	3	7	8	6	5	9	1
8	6	9	5	4	1	3	7	2
9	3	4	6	1	8	7	2	5
6	1	2	9	7	5	8	3	4
5	7	8	3	2	4	1	6	9

EASY - 183

8	4	9	1	7	5	6	3	2
5	7	6	2	9	3	1	4	8
2	1	3	6	8	4	7	5	9
4	2	5	7	3	8	9	6	1
6	3	7	5	1	9	8	2	4
1	9	8	4	6	2	3	7	5
3	8	2	9	4	7	5	1	6
7	6	4	8	5	1	2	9	3
9	5	1	3	2	6	4	8	7

EASY - 184

2	3	7	4	6	5	8	9	1
9	6	1	3	7	8	2	4	5
4	5	8	2	1	9	7	6	3
3	4	9	6	8	2	1	5	7
8	7	6	5	9	1	3	2	4
5	1	2	7	4	3	6	8	9
1	2	4	9	3	6	5	7	8
6	9	3	8	5	7	4	1	2
7	8	5	1	2	4	9	3	6

EASY - 185

3	8	6	4	9	5	7	1	2
5	1	9	2	8	7	6	3	4
2	4	7	1	6	3	8	5	9
6	7	1	3	4	9	5	2	8
8	9	5	6	1	2	3	4	7
4	3	2	7	5	8	9	6	1
7	5	4	8	2	6	1	9	3
1	6	3	9	7	4	2	8	5
9	2	8	5	3	1	4	7	6

EASY - 186

6	1	3	4	7	8	2	5	9
7	5	9	2	3	1	4	8	6
2	4	8	5	6	9	3	7	1
1	3	5	7	8	2	6	9	4
4	9	7	3	5	6	1	2	8
8	2	6	9	1	4	5	3	7
9	7	1	6	2	5	8	4	3
3	6	2	8	4	7	9	1	5
5	8	4	1	9	3	7	6	2

EASY - 187

4	2	3	9	6	5	7	8	1
1	7	9	3	8	2	6	5	4
6	5	8	4	7	1	9	3	2
7	8	5	2	9	3	4	1	6
3	9	6	8	1	4	2	7	5
2	1	4	7	5	6	8	9	3
8	3	1	6	2	7	5	4	9
5	6	7	1	4	9	3	2	8
9	4	2	5	3	8	1	6	7

EASY - 188

6	9	1	4	2	3	8	7	5
3	4	5	9	7	8	6	2	1
8	7	2	1	6	5	9	3	4
1	3	9	6	4	7	2	5	8
4	8	7	5	9	2	3	1	6
2	5	6	3	8	1	7	4	9
5	2	4	8	3	6	1	9	7
9	6	3	7	1	4	5	8	2
7	1	8	2	5	9	4	6	3

EASY - 189

7	9	5	2	8	6	1	3	4
8	4	3	1	5	7	9	6	2
1	2	6	3	4	9	8	5	7
6	7	9	8	2	5	4	1	3
3	1	8	7	6	4	2	9	5
2	5	4	9	1	3	6	7	8
5	6	1	4	7	2	3	8	9
4	3	7	6	9	8	5	2	1
9	8	2	5	3	1	7	4	6

EASY - 190

2	5	7	1	4	3	6	8	9
4	6	1	9	7	8	3	2	5
8	9	3	2	5	6	1	4	7
7	3	5	8	1	2	4	9	6
1	4	8	5	6	9	2	7	3
9	2	6	4	3	7	8	5	1
6	8	2	3	9	5	7	1	4
3	1	9	7	8	4	5	6	2
5	7	4	6	2	1	9	3	8

EASY - 191

9	7	5	8	1	3	6	2	4
3	1	4	5	2	6	9	8	7
6	8	2	4	7	9	1	3	5
8	9	6	3	5	7	2	4	1
5	4	1	9	6	2	8	7	3
7	2	3	1	4	8	5	6	9
4	6	8	7	9	5	3	1	2
2	5	7	6	3	1	4	9	8
1	3	9	2	8	4	7	5	6

EASY - 192

1	9	4	5	6	3	7	8	2
3	2	8	7	9	4	5	6	1
5	6	7	2	1	8	3	9	4
6	8	1	4	3	9	2	7	5
4	7	3	8	2	5	6	1	9
2	5	9	6	7	1	8	4	3
8	3	5	9	4	7	1	2	6
7	4	2	1	5	6	9	3	8
9	1	6	3	8	2	4	5	7

EASY - 193								
5	6	1	7	3	2	8	9	4
4	8	3	9	1	5	6	2	7
7	2	9	4	8	6	1	5	3
2	3	6	8	5	9	4	7	1
8	1	5	2	4	7	3	6	9
9	4	7	3	6	1	2	8	5
3	7	8	6	9	4	5	1	2
6	5	2	1	7	3	9	4	8
1	9	4	5	2	8	7	3	6

EASY - 194								
8	6	1	2	7	4	5	9	3
3	9	7	8	6	5	4	1	2
2	5	4	9	3	1	8	6	7
7	2	9	3	5	6	1	8	4
6	3	8	4	1	2	7	5	9
1	4	5	7	8	9	3	2	6
9	7	3	1	2	8	6	4	5
4	8	6	5	9	3	2	7	1
5	1	2	6	4	7	9	3	8

EASY - 195								
4	6	2	1	9	7	5	8	3
9	8	3	2	5	4	1	7	6
7	5	1	6	3	8	9	2	4
3	7	5	9	2	1	6	4	8
1	4	6	8	7	5	2	3	9
2	9	8	4	6	3	7	5	1
8	2	4	5	1	9	3	6	7
6	1	7	3	4	2	8	9	5
5	3	9	7	8	6	4	1	2

EASY - 196								
1	9	2	8	4	3	5	7	6
6	5	7	1	9	2	4	3	8
4	8	3	5	7	6	1	9	2
7	2	8	3	6	4	9	5	1
5	6	9	7	8	1	3	2	4
3	4	1	2	5	9	6	8	7
2	1	5	4	3	7	8	6	9
9	3	4	6	2	8	7	1	5
8	7	6	9	1	5	2	4	3

EASY - 197								
6	1	5	2	3	8	9	4	7
7	4	9	6	1	5	3	2	8
2	3	8	7	9	4	5	1	6
9	2	6	1	4	7	8	3	5
1	5	3	9	8	6	2	7	4
4	8	7	5	2	3	6	9	1
5	6	2	3	7	1	4	8	9
8	9	1	4	5	2	7	6	3
3	7	4	8	6	9	1	5	2

EASY - 198								
5	7	6	9	4	2	1	8	[illegible]
2	4	1	3	7	8	5	9	[illegible]
3	9	8	5	6	1	4	7	[illegible]
1	3	9	4	5	7	6	2	[illegible]
8	5	7	6	2	3	9	1	[illegible]
4	6	2	1	8	9	7	3	[illegible]
6	1	3	8	9	5	2	4	[illegible]
9	2	5	7	3	4	8	6	[illegible]
7	8	4	2	1	6	3	5	[illegible]

EASY - 199								
4	7	3	2	5	6	1	8	9
2	8	1	4	3	9	5	7	6
6	9	5	7	8	1	4	2	3
1	3	4	9	2	5	8	6	7
8	2	6	1	7	3	9	4	5
7	5	9	8	6	4	3	1	2
9	6	8	5	4	2	7	3	1
3	1	7	6	9	8	2	5	4
5	4	2	3	1	7	6	9	8

EASY - 200								
1	9	8	4	6	3	2	7	5
6	2	7	8	1	5	3	9	4
3	4	5	9	7	2	8	1	6
2	3	1	5	9	7	4	6	8
7	8	9	6	3	4	1	5	2
5	6	4	1	2	8	9	3	7
8	1	6	2	5	9	7	4	3
4	5	3	7	8	1	6	2	9
9	7	2	3	4	6	5	8	1

SUDOKU
SOLUTIONS
INTERMEDIATE - {1 -300}

INTERMEDIATE - 01

4	9	2	1	6	8	3	5	7
8	1	7	9	5	3	6	4	2
5	6	3	7	2	4	1	8	9
1	4	5	6	8	7	2	9	3
7	3	8	2	4	9	5	1	6
6	2	9	3	1	5	4	7	8
3	5	4	8	7	6	9	2	1
2	7	6	5	9	1	8	3	4
9	8	1	4	3	2	7	6	5

INTERMEDIATE - 02

6	8	4	9	2	7	3	1	5
3	9	2	4	5	1	6	7	8
7	1	5	6	8	3	9	4	2
4	2	7	1	3	6	8	5	9
9	5	3	2	4	8	7	6	1
1	6	8	5	7	9	2	3	4
5	7	1	8	6	2	4	9	3
2	4	6	3	9	5	1	8	7
8	3	9	7	1	4	5	2	6

INTERMEDIATE - 03

3	1	5	7	9	6	4	2	8
4	2	6	1	8	5	7	3	9
9	8	7	3	4	2	1	6	5
2	3	1	4	7	8	5	9	6
5	4	8	6	2	9	3	7	1
6	7	9	5	1	3	8	4	2
7	5	2	9	3	1	6	8	4
1	9	3	8	6	4	2	5	7
8	6	4	2	5	7	9	1	3

INTERMEDIATE - 04

5	4	6	2	9	8	1	7	3
2	3	7	1	6	5	9	4	8
1	8	9	4	3	7	2	6	5
6	9	8	3	7	2	5	1	4
3	2	4	5	8	1	7	9	6
7	1	5	9	4	6	3	8	2
8	6	2	7	1	3	4	5	9
9	7	3	6	5	4	8	2	1
4	5	1	8	2	9	6	3	7

INTERMEDIATE - 05

8	6	2	5	3	7	4	1	9
1	9	5	6	4	2	7	3	8
3	4	7	1	9	8	6	2	5
2	1	8	7	5	9	3	6	4
6	3	9	2	8	4	5	7	1
5	7	4	3	6	1	9	8	2
7	2	6	9	1	5	8	4	3
9	8	3	4	2	6	1	5	7
4	5	1	8	7	3	2	9	6

INTERMEDIATE - 06

9	2	5	1	7	8	4	3	6
7	3	4	6	5	2	1	9	8
1	8	6	9	4	3	2	7	5
3	1	2	5	9	6	7	8	4
4	6	8	3	1	7	9	5	2
5	7	9	8	2	4	3	6	1
2	4	3	7	6	5	8	1	9
6	9	7	4	8	1	5	2	3
8	5	1	2	3	9	6	4	7

INTERMEDIATE - 07

2	8	1	4	6	7	9	5	3
5	4	9	3	1	8	7	6	2
7	6	3	2	9	5	1	8	4
4	2	6	1	3	9	8	7	5
9	1	7	5	8	2	4	3	6
3	5	8	7	4	6	2	1	9
1	7	4	9	5	3	6	2	8
8	9	5	6	2	1	3	4	7
6	3	2	8	7	4	5	9	1

INTERMEDIATE - 08

2	8	6	4	5	1	3	7	9
5	7	4	9	6	3	1	8	2
3	1	9	8	2	7	4	6	5
4	3	2	6	1	9	8	5	7
9	6	7	3	8	5	2	4	1
1	5	8	7	4	2	6	9	3
6	2	3	5	7	4	9	1	8
7	4	1	2	9	8	5	3	6
8	9	5	1	3	6	7	2	4

INTERMEDIATE - 09

2	7	3	4	8	1	6	5	9
4	8	5	9	7	6	2	3	1
9	6	1	2	5	3	7	4	8
6	2	8	7	3	9	5	1	4
5	1	9	8	6	4	3	7	2
3	4	7	5	1	2	8	9	6
8	5	4	1	2	7	9	6	3
1	3	2	6	9	5	4	8	7
7	9	6	3	4	8	1	2	5

INTERMEDIATE - 10

7	8	5	4	6	9	3	2	1
3	6	1	2	7	5	8	9	4
9	2	4	8	3	1	5	6	7
5	7	3	9	1	4	6	8	2
4	9	6	7	8	2	1	3	5
2	1	8	3	5	6	4	7	9
6	4	7	5	2	8	9	1	3
1	3	9	6	4	7	2	5	8
8	5	2	1	9	3	7	4	6

INTERMEDIATE - 11

1	3	6	7	2	4	8	9	5
7	9	5	6	3	8	4	1	2
4	2	8	1	5	9	7	6	3
9	1	7	5	6	2	3	8	4
3	5	4	8	7	1	9	2	6
6	8	2	4	9	3	1	5	7
5	6	1	3	8	7	2	4	9
2	4	3	9	1	6	5	7	8
8	7	9	2	4	5	6	3	1

INTERMEDIATE - 12

2	3	8	4	5	9	7	6	1
5	1	7	8	2	6	3	9	4
6	9	4	7	3	1	2	8	5
9	7	5	2	1	8	6	4	3
1	6	2	5	4	3	9	7	8
4	8	3	6	9	7	1	5	2
7	2	1	9	8	4	5	3	6
8	5	6	3	7	2	4	1	9
3	4	9	1	6	5	8	2	7

INTERMEDIATE - 13

5	6	9	2	4	1	8	7	3
8	3	4	9	7	5	6	1	2
7	1	2	8	3	6	9	4	5
6	4	5	1	8	3	2	9	7
9	2	1	5	6	7	4	3	8
3	8	7	4	2	9	1	5	6
4	9	8	3	5	2	7	6	1
1	7	3	6	9	8	5	2	4
2	5	6	7	1	4	3	8	9

INTERMEDIATE - 14

5	1	2	9	8	4	3	6	7
8	9	6	7	5	3	2	4	1
3	4	7	1	6	2	9	8	5
7	3	5	6	1	8	4	9	2
9	6	4	2	3	5	1	7	8
1	2	8	4	9	7	6	5	3
6	7	3	5	2	9	8	1	4
2	5	9	8	4	1	7	3	6
4	8	1	3	7	6	5	2	9

INTERMEDIATE - 15

5	7	6	1	9	8	2	4	3
3	9	2	4	7	6	5	1	8
4	8	1	5	3	2	6	9	7
6	3	9	7	5	1	4	8	2
8	5	4	3	2	9	1	7	6
1	2	7	8	6	4	3	5	9
7	4	5	6	8	3	9	2	1
2	1	3	9	4	7	8	6	5
9	6	8	2	1	5	7	3	4

INTERMEDIATE - 16

8	4	7	1	6	3	5	2	9
6	1	2	9	5	4	7	3	8
9	5	3	2	8	7	1	6	4
1	8	4	3	7	9	6	5	2
3	9	6	4	2	5	8	1	7
7	2	5	8	1	6	4	9	3
4	3	1	5	9	8	2	7	6
5	6	8	7	3	2	9	4	1
2	7	9	6	4	1	3	8	5

INTERMEDIATE - 17

4	9	2	5	6	7	1	8	3
7	8	3	9	2	1	5	6	4
6	5	1	3	8	4	7	9	2
2	7	6	4	5	8	3	1	9
3	4	5	7	1	9	8	2	6
8	1	9	6	3	2	4	7	5
1	2	4	8	9	5	6	3	7
5	6	8	2	7	3	9	4	1
9	3	7	1	4	6	2	5	8

INTERMEDIATE - 18

3	7	5	8	6	4	2	9	1
2	9	4	7	5	1	8	6	3
8	6	1	3	9	2	5	4	7
1	5	8	6	2	3	4	7	9
7	4	6	5	8	9	3	1	2
9	3	2	4	1	7	6	8	5
6	1	9	2	3	8	7	5	4
4	8	3	1	7	5	9	2	6
5	2	7	9	4	6	1	3	8

INTERMEDIATE - 19

2	5	7	9	6	1	3	4	8
3	6	8	2	7	4	5	9	1
1	4	9	8	3	5	6	7	2
5	9	3	7	2	8	4	1	6
8	1	4	5	9	6	7	2	3
7	2	6	1	4	3	9	8	5
6	7	1	3	8	9	2	5	4
4	8	2	6	5	7	1	3	9
9	3	5	4	1	2	8	6	7

INTERMEDIATE - 20

5	1	2	7	9	3	4	6	8
6	9	3	5	4	8	7	2	1
7	8	4	2	6	1	9	5	3
1	7	6	4	8	2	3	9	5
4	5	9	1	3	6	2	8	7
3	2	8	9	7	5	6	1	4
8	3	7	6	1	9	5	4	2
9	4	5	8	2	7	1	3	6
2	6	1	3	5	4	8	7	9

INTERMEDIATE - 21

8	2	7	4	3	5	6	1	9
3	6	9	1	2	7	4	5	8
4	5	1	8	6	9	7	2	3
9	7	4	6	1	3	5	8	2
2	8	3	7	5	4	9	6	1
6	1	5	2	9	8	3	7	4
7	4	6	3	8	2	1	9	5
1	9	8	5	4	6	2	3	7
5	3	2	9	7	1	8	4	6

INTERMEDIATE - 22

7	5	2	1	8	3	4	9	6
3	9	1	2	4	6	5	7	8
4	8	6	5	9	7	1	2	3
9	4	3	8	1	5	7	6	2
1	6	8	9	7	2	3	5	4
5	2	7	6	3	4	8	1	9
6	3	4	7	2	1	9	8	5
8	7	5	3	6	9	2	4	1
2	1	9	4	5	8	6	3	7

INTERMEDIATE - 23

3	6	7	1	5	8	2	9	4
4	9	8	2	6	3	7	5	1
2	5	1	7	4	9	8	6	3
7	8	4	9	1	6	3	2	5
5	1	3	4	8	2	6	7	9
6	2	9	5	3	7	4	1	8
9	4	6	8	7	5	1	3	2
1	7	2	3	9	4	5	8	6
8	3	5	6	2	1	9	4	7

INTERMEDIATE - 24

3	5	6	1	4	7	9	8	2
1	7	8	2	9	3	4	6	5
9	2	4	5	6	8	7	1	3
7	8	3	4	5	9	1	2	6
4	6	2	3	8	1	5	9	7
5	9	1	7	2	6	3	4	8
2	1	7	8	3	4	6	5	9
8	4	9	6	7	5	2	3	1
6	3	5	9	1	2	8	7	4

INTERMEDIATE - 25

1	7	6	4	9	2	3	5	8
2	9	5	3	6	8	1	7	4
8	4	3	1	5	7	9	2	6
4	8	1	7	2	9	5	6	3
6	5	9	8	3	1	2	4	7
7	3	2	5	4	6	8	1	9
9	6	7	2	8	5	4	3	1
5	1	4	9	7	3	6	8	2
3	2	8	6	1	4	7	9	5

INTERMEDIATE - 26

6	1	2	7	5	8	9	3	4
4	9	5	3	6	1	2	7	8
7	8	3	9	2	4	1	5	6
1	2	8	5	3	9	4	6	7
3	7	6	4	8	2	5	1	9
5	4	9	1	7	6	3	8	2
8	6	4	2	1	5	7	9	3
2	3	1	6	9	7	8	4	5
9	5	7	8	4	3	6	2	1

INTERMEDIATE - 27

8	5	1	3	4	9	2	7	6
3	2	6	7	1	5	4	8	9
4	7	9	8	2	6	5	1	3
1	3	4	5	9	8	6	2	7
9	8	7	2	6	1	3	5	4
5	6	2	4	3	7	8	9	1
6	1	5	9	8	4	7	3	2
2	9	8	6	7	3	1	4	5
7	4	3	1	5	2	9	6	8

INTERMEDIATE - 28

8	6	9	5	4	2	1	3	7
5	4	7	1	6	3	8	2	9
3	2	1	9	7	8	6	4	5
4	8	5	3	2	7	9	6	1
1	3	2	6	5	9	4	7	8
7	9	6	8	1	4	3	5	2
9	5	8	7	3	6	2	1	4
6	1	4	2	8	5	7	9	3
2	7	3	4	9	1	5	8	6

INTERMEDIATE - 29

3	8	4	2	7	1	6	5	9
7	6	9	8	3	5	1	2	4
5	2	1	9	4	6	3	7	8
2	1	5	7	6	4	8	9	3
4	7	8	3	9	2	5	6	1
6	9	3	5	1	8	2	4	7
8	4	6	1	5	7	9	3	2
9	5	2	4	8	3	7	1	6
1	3	7	6	2	9	4	8	5

INTERMEDIATE - 30

5	8	6	9	7	3	1	4	2
4	9	7	5	1	2	8	6	3
3	1	2	8	6	4	5	7	9
2	5	9	6	3	7	4	8	1
6	3	1	4	8	5	9	2	7
7	4	8	2	9	1	3	5	6
1	6	4	3	2	8	7	9	5
9	7	5	1	4	6	2	3	8
8	2	3	7	5	9	6	1	4

INTERMEDIATE - 31

7	1	8	2	4	5	6	3	9
6	3	2	7	8	9	5	4	1
4	9	5	3	1	6	8	2	7
1	6	4	8	7	2	9	5	3
2	8	9	5	6	3	1	7	4
5	7	3	1	9	4	2	6	8
9	4	1	6	5	7	3	8	2
3	5	7	9	2	8	4	1	6
8	2	6	4	3	1	7	9	5

INTERMEDIATE - 32

3	9	4	1	6	5	7	2	8
2	5	6	3	7	8	1	4	9
8	1	7	2	4	9	6	3	5
7	8	3	9	5	2	4	1	6
4	2	9	6	1	3	5	8	7
1	6	5	7	8	4	3	9	2
5	3	1	8	2	7	9	6	4
6	7	2	4	9	1	8	5	3
9	4	8	5	3	6	2	7	1

INTERMEDIATE - 33

4	6	5	1	7	9	8	3	2
3	7	9	8	2	5	4	6	1
1	2	8	3	4	6	9	5	7
7	4	1	6	3	2	5	9	8
2	9	3	5	8	7	6	1	4
8	5	6	9	1	4	2	7	3
6	1	4	7	5	8	3	2	9
9	3	2	4	6	1	7	8	5
5	8	7	2	9	3	1	4	6

INTERMEDIATE - 34

1	4	9	3	7	5	8	2	6
8	6	3	1	2	4	9	5	7
5	7	2	9	8	6	4	1	3
6	3	4	5	9	2	1	7	8
9	5	7	4	1	8	6	3	2
2	1	8	6	3	7	5	9	4
7	2	1	8	4	9	3	6	5
4	9	5	2	6	3	7	8	1
3	8	6	7	5	1	2	4	9

INTERMEDIATE - 35

9	4	6	8	2	1	5	7	3
5	8	2	7	3	4	9	6	1
7	1	3	6	9	5	8	2	4
2	9	4	3	1	8	7	5	6
1	6	7	9	5	2	3	4	8
3	5	8	4	6	7	2	1	9
6	2	5	1	8	3	4	9	7
4	3	1	2	7	9	6	8	5
8	7	9	5	4	6	1	3	2

INTERMEDIATE - 36

8	9	5	4	6	1	3	2	7
1	2	3	8	7	5	9	6	4
4	6	7	9	3	2	1	8	5
5	8	6	7	4	9	2	3	1
2	3	1	6	5	8	7	4	9
7	4	9	2	1	3	8	5	6
9	5	4	3	8	7	6	1	2
3	1	2	5	9	6	4	7	8
6	7	8	1	2	4	5	9	3

INTERMEDIATE - 37

1	4	8	9	5	6	7	3	2
3	5	9	2	7	4	1	6	8
7	2	6	1	8	3	9	4	5
9	7	3	8	4	2	6	5	1
5	6	4	3	1	7	2	8	9
2	8	1	5	6	9	4	7	3
4	1	5	7	2	8	3	9	6
8	3	7	6	9	1	5	2	4
6	9	2	4	3	5	8	1	7

INTERMEDIATE - 38

9	1	5	7	2	4	3	8	6
2	3	8	1	9	6	7	5	4
4	6	7	5	8	3	2	9	1
8	4	1	9	7	5	6	2	3
5	9	2	3	6	1	8	4	7
3	7	6	2	4	8	5	1	9
1	8	9	6	3	2	4	7	5
6	5	4	8	1	7	9	3	2
7	2	3	4	5	9	1	6	8

INTERMEDIATE - 39

5	9	8	6	3	2	1	7	4
1	6	4	5	7	8	3	2	9
3	7	2	9	1	4	5	6	8
4	5	3	1	9	6	7	8	2
2	1	7	4	8	3	9	5	6
6	8	9	7	2	5	4	1	3
8	4	1	2	5	9	6	3	7
9	3	5	8	6	7	2	4	1
7	2	6	3	4	1	8	9	5

INTERMEDIATE - 40

9	2	1	7	8	5	6	4	3
3	7	4	9	6	1	2	8	5
8	6	5	3	4	2	1	9	7
2	8	3	6	1	9	7	5	4
1	4	7	5	2	8	9	3	6
5	9	6	4	3	7	8	1	2
7	5	2	8	9	3	4	6	1
6	1	9	2	5	4	3	7	8
4	3	8	1	7	6	5	2	9

INTERMEDIATE - 41

3	5	9	6	7	2	1	8	4
7	4	6	5	1	8	9	2	3
2	8	1	3	9	4	5	6	7
5	3	7	2	6	9	8	4	1
4	6	2	8	5	1	7	3	9
1	9	8	4	3	7	6	5	2
6	1	3	7	4	5	2	9	8
8	7	4	9	2	6	3	1	5
9	2	5	1	8	3	4	7	6

INTERMEDIATE - 42

3	7	2	9	4	6	8	1	5
6	1	5	8	3	7	9	4	2
8	9	4	2	1	5	3	6	7
1	2	8	4	6	9	7	5	3
5	3	9	1	7	2	4	8	6
4	6	7	5	8	3	2	9	1
7	8	1	6	2	4	5	3	9
9	4	3	7	5	1	6	2	8
2	5	6	3	9	8	1	7	4

INTERMEDIATE - 43

4	1	6	2	5	8	3	7	9
5	8	2	9	7	3	6	4	1
3	7	9	1	4	6	2	5	8
2	5	3	4	1	9	7	8	6
7	9	8	5	6	2	4	1	3
6	4	1	8	3	7	9	2	5
9	6	4	7	8	1	5	3	2
1	3	5	6	2	4	8	9	7
8	2	7	3	9	5	1	6	4

INTERMEDIATE - 44

3	9	5	1	8	6	4	7	2
4	6	8	5	2	7	1	3	9
2	1	7	4	3	9	5	8	6
9	5	3	7	1	4	6	2	8
6	7	4	2	5	8	9	1	3
8	2	1	6	9	3	7	4	5
1	4	9	3	6	2	8	5	7
5	8	2	9	7	1	3	6	4
7	3	6	8	4	5	2	9	1

INTERMEDIATE - 45

3	4	6	7	2	8	1	9	5
7	8	2	9	5	1	3	6	4
9	5	1	6	3	4	8	2	7
5	1	7	4	6	3	9	8	2
2	3	9	1	8	5	4	7	6
4	6	8	2	7	9	5	3	1
6	9	5	8	4	7	2	1	3
8	7	3	5	1	2	6	4	9
1	2	4	3	9	6	7	5	8

INTERMEDIATE - 46

9	8	4	6	3	5	2	1	7
7	2	3	1	9	4	8	6	5
6	1	5	2	7	8	9	3	4
5	3	2	4	6	1	7	8	9
4	7	6	9	8	2	1	5	3
8	9	1	3	5	7	6	4	2
2	4	8	7	1	3	5	9	6
3	5	9	8	2	6	4	7	1
1	6	7	5	4	9	3	2	8

INTERMEDIATE - 47

2	5	8	7	6	9	1	4	3
1	9	3	4	5	8	2	7	6
7	6	4	2	3	1	9	8	5
4	7	1	6	2	5	3	9	8
8	2	9	1	4	3	6	5	7
6	3	5	8	9	7	4	1	2
5	4	2	9	8	6	7	3	1
9	8	7	3	1	2	5	6	4
3	1	6	5	7	4	8	2	9

INTERMEDIATE - 48

9	1	4	5	2	3	7	8	6
3	6	7	9	1	8	5	4	2
5	2	8	6	4	7	9	3	1
7	9	2	1	8	6	3	5	4
1	8	3	4	9	5	2	6	7
4	5	6	3	7	2	8	1	9
2	3	5	7	6	4	1	9	8
8	4	1	2	5	9	6	7	3
6	7	9	8	3	1	4	2	5

INTERMEDIATE - 49

6	8	2	1	7	5	9	4	3
3	7	9	4	2	6	1	5	8
4	5	1	9	8	3	2	6	7
9	2	8	7	1	4	6	3	5
1	3	4	6	5	9	8	7	2
7	6	5	8	3	2	4	9	1
5	9	3	2	4	8	7	1	6
2	4	7	3	6	1	5	8	9
8	1	6	5	9	7	3	2	4

INTERMEDIATE - 50

3	8	1	5	6	2	7	4	9
5	9	2	8	7	4	1	3	6
7	4	6	3	9	1	5	2	8
9	6	5	7	2	8	3	1	4
4	3	7	1	5	6	9	8	2
1	2	8	4	3	9	6	5	7
6	1	9	2	8	5	4	7	3
8	5	3	9	4	7	2	6	1
2	7	4	6	1	3	8	9	5

INTERMEDIATE - 51

8	2	4	5	3	9	1	7	6
5	6	3	8	1	7	9	2	4
9	1	7	2	6	4	3	8	5
3	4	8	6	7	1	2	5	9
1	5	6	9	2	8	7	4	3
7	9	2	3	4	5	8	6	1
2	3	1	7	5	6	4	9	8
4	8	5	1	9	2	6	3	7
6	7	9	4	8	3	5	1	2

INTERMEDIATE - 52

1	6	2	3	4	7	5	9	8
4	5	9	2	8	6	3	7	1
3	7	8	1	5	9	2	4	6
5	8	7	6	9	2	4	1	3
2	1	4	7	3	5	6	8	9
6	9	3	8	1	4	7	2	5
8	2	5	9	7	3	1	6	4
7	4	1	5	6	8	9	3	2
9	3	6	4	2	1	8	5	7

INTERMEDIATE - 53

1	4	6	7	3	9	5	8	2
3	2	5	8	4	1	6	7	9
8	7	9	6	5	2	4	1	3
6	5	7	2	9	8	1	3	4
2	3	1	4	7	6	9	5	8
4	9	8	3	1	5	2	6	7
7	6	3	1	2	4	8	9	5
5	1	2	9	8	3	7	4	6
9	8	4	5	6	7	3	2	1

INTERMEDIATE - 54

5	9	6	7	4	3	2	8	1
4	8	2	9	1	5	3	6	7
3	1	7	2	8	6	5	4	9
8	3	4	1	5	2	9	7	6
7	6	5	4	3	9	8	1	2
9	2	1	6	7	8	4	3	5
6	4	3	5	9	1	7	2	8
2	7	9	8	6	4	1	5	3
1	5	8	3	2	7	6	9	4

INTERMEDIATE - 55

8	7	1	9	4	3	2	5	6
5	3	6	1	7	2	8	9	4
2	9	4	5	6	8	7	3	1
1	2	8	3	5	7	4	6	9
4	6	7	2	9	1	3	8	5
3	5	9	4	8	6	1	7	2
6	1	2	7	3	9	5	4	8
9	4	3	8	1	5	6	2	7
7	8	5	6	2	4	9	1	3

INTERMEDIATE - 56

7	5	8	9	4	2	3	6	1
9	6	3	7	8	1	5	2	4
2	1	4	6	5	3	7	9	8
1	8	2	3	7	9	6	4	5
6	4	5	1	2	8	9	3	7
3	7	9	4	6	5	1	8	2
4	3	6	2	1	7	8	5	9
8	9	1	5	3	4	2	7	6
5	2	7	8	9	6	4	1	3

INTERMEDIATE - 57

1	3	6	9	7	5	2	8	4
2	8	5	3	1	4	6	9	7
7	4	9	6	8	2	3	1	5
6	5	4	8	2	3	9	7	1
8	7	3	1	9	6	5	4	2
9	1	2	5	4	7	8	3	6
5	9	1	7	6	8	4	2	3
4	6	8	2	3	1	7	5	9
3	2	7	4	5	9	1	6	8

INTERMEDIATE - 58

3	1	4	9	2	7	5	8	6
6	5	2	8	3	1	7	4	9
9	7	8	5	4	6	2	1	3
1	8	3	6	5	2	9	7	4
7	2	5	1	9	4	6	3	8
4	6	9	3	7	8	1	2	5
5	9	1	7	8	3	4	6	2
2	3	6	4	1	9	8	5	7
8	4	7	2	6	5	3	9	1

INTERMEDIATE - 59

3	4	1	5	8	9	7	2	6
7	9	5	4	6	2	1	3	8
2	8	6	7	3	1	4	5	9
6	3	4	2	9	5	8	7	1
5	7	9	3	1	8	6	4	2
1	2	8	6	7	4	5	9	3
9	5	2	1	4	6	3	8	7
4	1	3	8	2	7	9	6	5
8	6	7	9	5	3	2	1	4

INTERMEDIATE - 60

5	2	4	3	8	7	6	9	1
8	9	6	2	1	5	3	7	4
7	3	1	6	9	4	8	5	2
1	7	8	9	2	3	4	6	5
6	4	3	7	5	1	9	2	8
2	5	9	4	6	8	7	1	3
9	6	5	8	4	2	1	3	7
4	1	7	5	3	9	2	8	6
3	8	2	1	7	6	5	4	9

INTERMEDIATE - 61

4	8	2	6	3	1	9	7	5
6	3	5	9	7	8	4	2	1
9	7	1	4	5	2	3	6	8
8	5	6	3	9	4	2	1	7
2	9	4	7	1	5	8	3	6
7	1	3	2	8	6	5	4	9
5	4	9	1	2	7	6	8	3
3	6	7	8	4	9	1	5	2
1	2	8	5	6	3	7	9	4

INTERMEDIATE - 62

8	7	4	9	2	1	3	5	6
6	1	9	7	5	3	2	8	4
2	3	5	4	8	6	1	7	9
3	6	8	2	1	9	5	4	7
5	4	7	6	3	8	9	1	2
9	2	1	5	7	4	6	3	8
7	8	6	1	9	5	4	2	3
1	9	3	8	4	2	7	6	5
4	5	2	3	6	7	8	9	1

INTERMEDIATE - 63

8	6	4	7	9	1	3	5	2
1	9	5	2	4	3	6	8	7
3	2	7	5	8	6	1	4	9
4	1	2	9	5	8	7	6	3
9	7	6	3	1	4	8	2	5
5	3	8	6	7	2	9	1	4
7	5	1	4	6	9	2	3	8
6	4	3	8	2	7	5	9	1
2	8	9	1	3	5	4	7	6

INTERMEDIATE - 64

8	3	9	2	1	6	5	4	7
5	1	4	3	8	7	6	2	9
7	6	2	5	9	4	3	1	8
1	2	8	6	4	3	7	9	5
3	9	5	7	2	1	4	8	6
6	4	7	8	5	9	1	3	2
2	5	3	1	7	8	9	6	4
9	8	6	4	3	5	2	7	1
4	7	1	9	6	2	8	5	3

INTERMEDIATE - 65

6	4	9	7	1	8	2	3	5
8	7	5	9	2	3	6	1	4
2	3	1	6	5	4	9	7	8
9	5	8	4	7	2	3	6	1
7	1	4	3	6	5	8	2	9
3	2	6	8	9	1	4	5	7
1	8	2	5	4	6	7	9	3
4	6	7	1	3	9	5	8	2
5	9	3	2	8	7	1	4	6

INTERMEDIATE - 66

7	1	3	6	4	5	8	2	9
5	8	9	7	2	3	4	6	1
2	6	4	8	9	1	3	5	7
3	9	1	5	7	6	2	4	8
8	7	5	4	1	2	6	9	3
6	4	2	9	3	8	1	7	5
4	2	8	1	5	9	7	3	6
1	5	7	3	6	4	9	8	2
9	3	6	2	8	7	5	1	4

INTERMEDIATE - 67

4	2	7	5	8	6	1	9	3
8	1	3	9	7	4	2	5	6
5	6	9	3	2	1	7	4	8
9	3	6	4	1	8	5	7	2
1	7	4	6	5	2	3	8	9
2	8	5	7	9	3	6	1	4
6	4	8	1	3	7	9	2	5
7	5	2	8	6	9	4	3	1
3	9	1	2	4	5	8	6	7

INTERMEDIATE - 68

2	6	4	3	5	7	8	1	9
9	8	7	4	1	2	3	5	6
1	5	3	6	9	8	7	2	4
3	9	5	2	8	6	1	4	7
7	2	8	9	4	1	5	6	3
6	4	1	7	3	5	2	9	8
8	1	6	5	7	9	4	3	2
4	7	9	1	2	3	6	8	5
5	3	2	8	6	4	9	7	1

INTERMEDIATE - 69

4	7	5	9	8	3	6	1	2
1	3	9	2	6	5	7	8	4
2	6	8	7	1	4	9	5	3
6	5	3	4	2	8	1	7	9
7	1	2	6	5	9	4	3	8
9	8	4	3	7	1	2	6	5
3	2	1	5	9	6	8	4	7
5	9	6	8	4	7	3	2	1
8	4	7	1	3	2	5	9	6

INTERMEDIATE - 70

5	1	6	4	9	8	7	3	2
9	8	3	7	2	6	4	1	5
7	4	2	1	5	3	6	9	8
4	5	9	6	3	1	2	8	7
6	3	1	2	8	7	5	4	9
2	7	8	5	4	9	1	6	3
3	2	4	9	6	5	8	7	1
1	9	5	8	7	4	3	2	6
8	6	7	3	1	2	9	5	4

INTERMEDIATE - 71

7	3	4	9	2	8	6	1	5
5	1	9	3	6	7	2	8	4
2	6	8	5	4	1	9	3	7
3	4	1	7	5	2	8	9	6
6	8	7	1	3	9	4	5	2
9	2	5	4	8	6	1	7	3
1	7	6	2	9	5	3	4	8
8	9	3	6	7	4	5	2	1
4	5	2	8	1	3	7	6	9

INTERMEDIATE - 72

5	1	8	3	2	7	9	6	4
7	4	2	6	9	8	5	1	3
6	3	9	4	1	5	2	8	7
3	9	4	5	8	2	1	7	6
8	6	1	7	3	9	4	5	2
2	5	7	1	4	6	8	3	9
1	7	6	9	5	4	3	2	8
4	2	5	8	7	3	6	9	1
9	8	3	2	6	1	7	4	5

INTERMEDIATE - 73

1	3	6	9	5	4	2	7	8
7	5	9	8	2	1	3	4	6
8	4	2	7	3	6	1	5	9
2	7	3	1	4	9	6	8	5
6	8	4	5	7	3	9	1	2
9	1	5	6	8	2	4	3	7
5	9	1	4	6	7	8	2	3
3	6	8	2	1	5	7	9	4
4	2	7	3	9	8	5	6	1

INTERMEDIATE - 74

1	6	9	8	5	4	2	3	7
8	2	4	3	7	6	9	5	1
5	3	7	1	9	2	4	6	8
2	5	3	9	1	8	6	7	4
6	7	8	4	2	3	1	9	5
9	4	1	5	6	7	3	8	2
3	8	2	7	4	9	5	1	6
7	1	6	2	3	5	8	4	9
4	9	5	6	8	1	7	2	3

INTERMEDIATE - 75

1	7	3	8	4	9	5	2	6
4	6	5	2	1	3	8	7	9
8	2	9	6	5	7	1	4	3
2	4	8	9	6	5	7	3	1
3	1	6	7	2	8	4	9	5
9	5	7	4	3	1	6	8	2
6	3	2	5	7	4	9	1	8
5	9	4	1	8	2	3	6	7
7	8	1	3	9	6	2	5	4

INTERMEDIATE - 76

1	6	2	3	9	8	4	7	5
5	3	8	4	7	6	1	2	9
9	4	7	1	2	5	8	3	6
3	2	9	8	6	7	5	4	1
4	7	6	5	1	3	2	9	8
8	1	5	9	4	2	7	6	3
2	9	4	6	5	1	3	8	7
6	5	3	7	8	4	9	1	2
7	8	1	2	3	9	6	5	4

INTERMEDIATE -77

4	7	6	1	2	8	5	9	3
1	5	2	7	9	3	4	6	8
3	9	8	6	5	4	7	2	1
2	8	5	9	1	6	3	4	7
6	1	7	4	3	5	2	8	9
9	3	4	8	7	2	6	1	5
7	6	3	2	8	1	9	5	4
5	4	1	3	6	9	8	7	2
8	2	9	5	4	7	1	3	6

INTERMEDIATE - 78

Sudoku Solution 78

2	8	9	7	4	1	6	5	3
6	1	5	8	3	9	2	7	4
3	7	4	2	5	6	1	9	8
7	6	1	5	2	4	8	3	9
9	4	3	1	8	7	5	6	2
8	5	2	6	9	3	4	1	7
1	9	6	4	7	8	3	2	5
5	3	8	9	1	2	7	4	6

INTERMEDIATE - 79

5	2	6	7	8	4	3	1	9
7	3	8	1	5	9	4	2	6
9	4	1	6	2	3	5	7	8
2	5	3	8	4	1	6	9	7
8	9	4	5	7	6	2	3	1
6	1	7	9	3	2	8	5	4
1	8	2	3	6	7	9	4	5
4	6	9	2	1	5	7	8	3
3	7	5	4	9	8	1	6	2

INTERMEDIATE - 80

2	4	8	9	5	6	3	1	7
6	3	9	8	7	1	2	4	5
5	1	7	4	3	2	8	9	6
1	5	6	7	4	8	9	3	2
4	7	3	2	6	9	5	8	1
9	8	2	3	1	5	7	6	4
3	2	1	6	9	7	4	5	8
7	6	4	5	8	3	1	2	9
8	9	5	1	2	4	6	7	3

INTERMEDIATE - 81

2	9	1	6	8	5	3	4	7
6	7	8	9	4	3	2	1	5
3	5	4	1	7	2	9	8	6
7	4	6	3	1	9	8	5	2
5	3	9	8	2	7	1	6	4
8	1	2	5	6	4	7	3	9
4	2	5	7	3	1	6	9	8
1	6	7	4	9	8	5	2	3
9	8	3	2	5	6	4	7	1

INTERMEDIATE - 82

8	9	7	5	2	6	3	4	1
2	1	4	9	8	3	5	7	6
5	6	3	4	1	7	2	8	9
1	3	9	8	5	4	7	6	2
4	5	8	6	7	2	1	9	3
6	7	2	1	3	9	8	5	4
3	8	6	2	9	5	4	1	7
7	4	1	3	6	8	9	2	5
9	2	5	7	4	1	6	3	8

INTERMEDIATE - 83

5	1	3	4	2	6	9	7	8
2	4	8	9	3	7	6	5	1
9	7	6	5	1	8	3	2	4
3	6	2	1	8	9	5	4	7
7	9	4	2	6	5	1	8	3
8	5	1	3	7	4	2	9	6
1	3	9	7	4	2	8	6	5
6	2	7	8	5	3	4	1	9
4	8	5	6	9	1	7	3	2

INTERMEDIATE - 84

6	4	9	1	5	7	2	8	3
7	2	3	4	6	8	5	9	1
5	1	8	9	3	2	7	4	6
8	6	2	7	9	5	1	3	4
9	7	1	3	8	4	6	2	5
3	5	4	2	1	6	8	7	9
4	8	5	6	7	3	9	1	2
2	9	6	8	4	1	3	5	7
1	3	7	5	2	9	4	6	8

INTERMEDIATE - 85

4	8	9	6	1	2	3	5	7
6	1	7	5	3	4	9	2	8
2	5	3	7	8	9	1	4	6
5	3	2	9	6	1	7	8	4
7	9	6	8	4	5	2	1	3
1	4	8	3	2	7	6	9	5
9	7	4	2	5	6	8	3	1
3	2	5	1	7	8	4	6	9
8	6	1	4	9	3	5	7	2

INTERMEDIATE - 86

1	3	9	6	2	5	7	4	8
7	5	6	4	8	3	2	9	1
8	4	2	1	9	7	6	3	5
2	6	8	3	5	9	4	1	7
3	9	7	8	4	1	5	6	2
4	1	5	2	7	6	3	8	9
9	2	1	5	3	4	8	7	6
6	8	3	7	1	2	9	5	4
5	7	4	9	6	8	1	2	3

INTERMEDIATE - 87

5	8	9	7	3	6	4	1	2
3	4	2	1	8	5	9	7	6
1	6	7	4	2	9	8	5	3
2	9	8	6	1	3	5	4	7
6	1	5	9	4	7	3	2	8
4	7	3	8	5	2	1	6	9
9	2	1	5	7	8	6	3	4
7	5	6	3	9	4	2	8	1
8	3	4	2	6	1	7	9	5

INTERMEDIATE - 88

9	3	6	1	5	7	4	2	8
2	5	1	8	6	4	3	7	9
8	7	4	2	9	3	6	1	5
7	9	5	3	2	8	1	4	6
1	4	8	9	7	6	2	5	3
6	2	3	4	1	5	8	9	7
4	6	7	5	3	1	9	8	2
5	1	2	6	8	9	7	3	4
3	8	9	7	4	2	5	6	1

INTERMEDIATE - 89

1	4	2	7	5	9	8	6	3
5	6	9	2	3	8	7	1	4
7	8	3	4	6	1	2	9	5
2	3	6	5	8	7	9	4	1
9	5	7	3	1	4	6	2	8
8	1	4	6	9	2	3	5	7
3	2	1	8	4	6	5	7	9
6	9	5	1	7	3	4	8	2
4	7	8	9	2	5	1	3	6

INTERMEDIATE - 90

5	9	8	6	2	3	4	7	1
2	3	4	7	1	8	6	5	9
1	6	7	9	4	5	8	2	3
9	7	1	5	8	4	2	3	6
4	2	3	1	9	6	5	8	7
8	5	6	2	3	7	1	9	4
6	1	9	3	5	2	7	4	8
7	4	5	8	6	9	3	1	2
3	8	2	4	7	1	9	6	5

INTERMEDIATE - 91

5	7	6	4	8	3	2	9	1
4	8	9	1	5	2	7	6	3
1	2	3	7	9	6	4	5	8
7	4	1	8	3	9	6	2	5
3	9	2	6	4	5	1	8	7
8	6	5	2	1	7	3	4	9
6	1	4	9	7	8	5	3	2
9	3	7	5	2	4	8	1	6
2	5	8	3	6	1	9	7	4

INTERMEDIATE - 92

5	1	7	4	2	9	3	6	8
2	6	9	7	8	3	4	5	1
4	8	3	6	1	5	9	7	2
1	9	5	2	7	6	8	4	3
6	2	8	9	3	4	5	1	7
7	3	4	8	5	1	2	9	6
9	5	2	1	6	8	7	3	4
3	7	6	5	4	2	1	8	9
8	4	1	3	9	7	6	2	5

INTERMEDIATE - 93

6	7	5	9	3	4	2	8	1
8	9	2	5	7	1	3	4	6
3	4	1	6	8	2	5	9	7
4	1	3	8	9	6	7	5	2
5	2	8	7	1	3	9	6	4
7	6	9	2	4	5	1	3	8
1	3	7	4	6	9	8	2	5
9	5	6	1	2	8	4	7	3
2	8	4	3	5	7	6	1	9

INTERMEDIATE - 94

9	5	8	4	3	1	2	6	7
7	4	2	5	6	8	3	9	1
3	6	1	9	7	2	8	4	5
5	7	9	8	1	3	6	2	4
8	1	6	2	4	9	5	7	3
2	3	4	6	5	7	9	1	8
1	8	3	7	2	6	4	5	9
6	9	5	1	8	4	7	3	2
4	2	7	3	9	5	1	8	6

INTERMEDIATE - 95

2	9	7	4	8	6	5	3	1
4	5	8	1	9	3	2	6	7
1	3	6	5	7	2	8	4	9
8	4	5	7	2	9	6	1	3
3	1	9	6	4	5	7	8	2
6	7	2	3	1	8	4	9	5
7	2	4	9	6	1	3	5	8
9	8	3	2	5	4	1	7	6
5	6	1	8	3	7	9	2	4

INTERMEDIATE - 96

9	6	2	5	7	8	1	4	3
5	4	3	1	9	2	6	8	7
1	7	8	6	3	4	9	5	2
2	3	4	7	5	9	8	1	6
7	8	9	4	1	6	3	2	5
6	5	1	2	8	3	4	7	9
3	1	6	8	2	5	7	9	4
4	2	7	9	6	1	5	3	8
8	9	5	3	4	7	2	6	1

INTERMEDIATE - 97

7	9	4	5	8	2	3	6	1
5	8	3	6	1	9	4	2	7
6	1	2	4	7	3	8	9	5
3	6	9	1	2	4	5	7	8
4	5	8	3	9	7	2	1	6
2	7	1	8	6	5	9	3	4
1	3	7	9	4	8	6	5	2
8	2	5	7	3	6	1	4	9
9	4	6	2	5	1	7	8	3

INTERMEDIATE - 98

2	1	8	7	5	9	3	4	6
6	9	5	3	2	4	7	8	1
7	4	3	8	6	1	9	2	5
8	7	6	9	4	3	5	1	2
9	2	4	1	8	5	6	7	3
5	3	1	2	7	6	8	9	4
3	5	2	4	9	7	1	6	8
1	8	7	6	3	2	4	5	9
4	6	9	5	1	8	2	3	7

INTERMEDIATE - 99

4	2	7	3	9	1	8	5	6
8	3	9	5	7	6	2	4	1
5	1	6	4	2	8	3	7	9
7	9	3	8	5	2	6	1	4
1	8	5	6	4	7	9	2	3
6	4	2	9	1	3	7	8	5
2	6	8	1	3	5	4	9	7
3	5	4	7	8	9	1	6	2
9	7	1	2	6	4	5	3	8

INTERMEDIATE - 100

3	7	8	4	6	2	1	9	5
1	4	6	9	7	5	2	3	8
5	2	9	1	8	3	7	4	6
8	5	1	3	4	9	6	7	2
6	9	2	8	1	7	3	5	4
7	3	4	2	5	6	9	8	1
2	1	5	7	3	8	4	6	9
4	6	3	5	9	1	8	2	7
9	8	7	6	2	4	5	1	3

INTERMEDIATE - 101

1	9	7	5	2	8	6	4	3
5	2	8	3	6	4	9	7	1
4	3	6	9	1	7	2	8	5
3	4	2	8	7	6	5	1	9
7	6	9	2	5	1	4	3	8
8	5	1	4	9	3	7	6	2
9	7	3	1	4	2	8	5	6
6	8	5	7	3	9	1	2	4
2	1	4	6	8	5	3	9	7

INTERMEDIATE - 102

6	1	9	4	2	8	3	7
3	4	2	5	7	1	6	8
7	8	5	9	6	3	1	4
1	9	8	6	3	4	5	2
2	6	4	8	5	7	9	1
5	7	3	2	1	9	8	6
8	3	6	7	9	2	4	5
9	5	7	1	4	6	2	3
4	2	1	3	8	5	7	9

INTERMEDIATE - 103

9	6	3	4	1	2	5	8	7
7	8	2	6	5	9	3	1	4
5	1	4	7	3	8	6	2	9
8	3	6	5	7	1	9	4	2
1	2	5	9	6	4	8	7	3
4	7	9	8	2	3	1	6	5
2	4	8	1	9	5	7	3	6
3	9	7	2	8	6	4	5	1
6	5	1	3	4	7	2	9	8

INTERMEDIATE - 104

3	9	1	7	4	6	5	8	2
4	5	7	8	2	9	1	3	6
6	2	8	5	3	1	4	7	9
5	6	4	9	1	3	8	2	7
8	3	2	4	7	5	9	6	1
7	1	9	2	6	8	3	4	5
9	7	6	3	5	4	2	1	8
1	8	3	6	9	2	7	5	4
2	4	5	1	8	7	6	9	3

INTERMEDIATE - 105

9	7	1	4	5	3	2	8	6
5	8	6	2	1	7	9	3	4
4	2	3	6	8	9	5	1	7
8	5	4	9	7	6	3	2	1
3	6	9	1	4	2	8	7	5
2	1	7	5	3	8	6	4	9
6	3	5	7	2	1	4	9	8
1	9	2	8	6	4	7	5	3
7	4	8	3	9	5	1	6	2

INTERMEDIATE - 106

5	7	8	2	4	9	1	6	3
3	2	4	7	1	6	8	9	5
9	6	1	3	8	5	7	2	4
7	8	6	4	9	1	5	3	2
2	4	9	5	3	7	6	1	8
1	5	3	8	6	2	9	4	7
4	9	7	1	5	3	2	8	6
6	3	5	9	2	8	4	7	1
8	1	2	6	7	4	3	5	9

INTERMEDIATE - 107

3	5	9	7	1	8	6	4	2
1	8	7	6	2	4	3	5	9
2	4	6	5	3	9	1	8	7
8	9	1	4	5	7	2	3	6
6	7	5	3	9	2	4	1	8
4	3	2	1	8	6	9	7	5
7	1	3	2	6	5	8	9	4
5	6	8	9	4	1	7	2	3
9	2	4	8	7	3	5	6	1

INTERMEDIATE - 108

5	6	9	4	3	2	8	1
3	4	7	1	6	8	9	2
1	2	8	9	5	7	3	6
6	3	1	2	8	4	5	7
4	9	2	5	7	1	6	8
8	7	5	3	9	6	1	4
2	5	6	7	1	3	4	9
9	1	4	8	2	5	7	3
7	8	3	6	4	9	2	5

INTERMEDIATE - 109

3	5	4	9	6	1	2	7	8
1	9	7	8	2	5	3	6	4
6	2	8	4	7	3	9	5	1
7	6	9	2	3	8	1	4	5
2	4	1	5	9	6	7	8	3
8	3	5	1	4	7	6	2	9
9	8	3	6	5	2	4	1	7
4	1	6	7	8	9	5	3	2
5	7	2	3	1	4	8	9	6

INTERMEDIATE - 110

9	7	1	8	2	3	6	5	4
3	5	4	9	6	7	8	1	2
6	2	8	4	5	1	7	3	9
1	8	9	7	4	2	5	6	3
7	3	6	5	1	9	4	2	8
5	4	2	6	3	8	1	9	7
4	9	3	1	7	5	2	8	6
2	6	5	3	8	4	9	7	1
8	1	7	2	9	6	3	4	5

INTERMEDIATE - 111

1	8	6	4	9	7	3	2	5
3	4	5	1	2	8	9	7	6
2	7	9	6	5	3	8	1	4
7	6	2	9	4	5	1	3	8
9	1	8	2	3	6	4	5	7
5	3	4	8	7	1	6	9	2
8	5	1	7	6	9	2	4	3
4	9	7	3	8	2	5	6	1
6	2	3	5	1	4	7	8	9

INTERMEDIATE - 112

8	3	2	7	9	1	6	5	4
7	6	9	5	3	4	1	8	2
5	4	1	8	2	6	9	3	7
4	7	8	6	1	9	5	2	3
6	9	3	2	7	5	8	4	1
1	2	5	3	4	8	7	9	6
9	8	7	4	6	3	2	1	5
2	5	4	1	8	7	3	6	9
3	1	6	9	5	2	4	7	8

INTERMEDIATE - 113

1	6	3	2	9	5	4	8	7
7	8	2	4	1	6	5	9	3
5	4	9	3	7	8	1	6	2
4	1	7	5	6	9	3	2	8
2	5	8	7	4	3	6	1	9
9	3	6	1	8	2	7	4	5
3	7	4	9	2	1	8	5	6
6	2	1	8	5	7	9	3	4
8	9	5	6	3	4	2	7	1

INTERMEDIATE - 114

8	9	1	5	4	3	7	2
3	2	7	1	6	8	9	4
4	5	6	7	9	2	3	8
6	4	9	2	5	7	1	3
2	7	8	6	3	1	4	5
1	3	5	4	8	9	6	7
7	8	2	9	1	4	5	6
9	6	4	3	2	5	8	1
5	1	3	8	7	6	2	9

INTERMEDIATE - 115

1	7	2	4	5	8	6	9	3
8	9	6	7	2	3	4	5	1
4	5	3	6	9	1	8	2	7
2	3	8	5	6	9	7	1	4
5	6	4	2	1	7	3	8	9
9	1	7	8	3	4	5	6	2
3	4	9	1	8	5	2	7	6
7	2	5	9	4	6	1	3	8
6	8	1	3	7	2	9	4	5

INTERMEDIATE - 116

3	9	6	8	5	4	7	1	2
8	4	7	3	2	1	5	9	6
1	2	5	7	6	9	3	8	4
6	8	2	9	3	5	4	7	1
7	3	1	4	8	6	9	2	5
4	5	9	2	1	7	6	3	8
9	6	3	1	4	8	2	5	7
5	7	8	6	9	2	1	4	3
2	1	4	5	7	3	8	6	9

INTERMEDIATE - 117

8	3	7	5	6	4	1	2	9
4	2	6	7	1	9	3	8	5
1	5	9	2	3	8	7	6	4
6	8	5	1	7	3	4	9	2
3	9	1	8	4	2	6	5	7
2	7	4	9	5	6	8	1	3
5	6	8	3	2	7	9	4	1
7	4	2	6	9	1	5	3	8
9	1	3	4	8	5	2	7	6

INTERMEDIATE - 118

2	7	4	9	6	5	3	8	1
1	8	6	4	3	2	5	9	7
9	5	3	7	1	8	4	6	2
4	9	8	1	2	6	7	3	5
3	1	7	5	9	4	8	2	6
5	6	2	3	8	7	9	1	4
6	2	5	8	7	3	1	4	9
8	4	9	6	5	1	2	7	3
7	3	1	2	4	9	6	5	8

INTERMEDIATE - 119

8	3	5	7	6	4	2	9	1
2	1	6	3	9	8	7	4	5
9	7	4	5	1	2	3	8	6
5	2	9	8	3	6	1	7	4
6	8	1	9	4	7	5	3	2
7	4	3	2	5	1	8	6	9
1	6	8	4	2	3	9	5	7
3	5	2	6	7	9	4	1	8
4	9	7	1	8	5	6	2	3

INTERMEDIATE - 120

2	1	6	3	4	7	9	8
4	7	3	9	8	5	2	6
8	9	5	1	6	2	4	7
6	3	1	2	7	8	5	9
9	2	7	5	1	4	6	3
5	8	4	6	3	9	7	1
7	5	2	8	9	1	3	4
1	6	9	4	5	3	8	2
3	4	8	7	2	6	1	5

INTERMEDIATE - 121

9	6	1	5	3	2	8	7	4
2	4	3	9	7	8	6	1	5
8	5	7	4	6	1	2	3	9
7	1	4	8	5	6	3	9	2
3	8	6	2	4	9	1	5	7
5	9	2	3	1	7	4	6	8
4	3	8	6	9	5	7	2	1
6	7	9	1	2	4	5	8	3
1	2	5	7	8	3	9	4	6

INTERMEDIATE - 122

7	5	3	9	8	1	4	6	2
2	1	9	6	7	4	8	3	5
8	4	6	5	3	2	7	9	1
9	7	5	1	4	3	2	8	6
1	8	2	7	6	9	3	5	4
6	3	4	8	2	5	1	7	9
4	6	1	3	5	7	9	2	8
5	9	7	2	1	8	6	4	3
3	2	8	4	9	6	5	1	7

INTERMEDIATE - 123

1	2	3	6	5	8	4	7	9
9	4	8	3	1	7	6	5	2
5	6	7	4	9	2	1	3	8
6	8	1	9	2	3	7	4	5
7	5	4	1	8	6	2	9	3
2	3	9	5	7	4	8	6	1
4	1	6	8	3	9	5	2	7
3	7	5	2	4	1	9	8	6
8	9	2	7	6	5	3	1	4

INTERMEDIATE - 124

8	1	2	3	4	5	6	9	7
4	9	3	2	6	7	1	8	5
6	5	7	8	1	9	4	2	3
3	2	5	1	8	6	7	4	9
7	4	6	5	9	2	3	1	8
1	8	9	4	7	3	5	6	2
2	6	8	7	3	4	9	5	1
5	7	4	9	2	1	8	3	6
9	3	1	6	5	8	2	7	4

INTERMEDIATE -125

7	6	9	4	5	2	3	8	1
8	5	1	9	7	3	4	6	2
4	2	3	8	6	1	7	5	9
6	3	2	1	9	7	5	4	8
1	9	8	3	4	5	2	7	6
5	4	7	2	8	6	1	9	3
2	8	4	7	3	9	6	1	5
9	1	6	5	2	4	8	3	7
3	7	5	6	1	8	9	2	4

INTERMEDIATE - 126

3	7	8	2	5	1	6	4
6	5	2	7	9	4	3	8
1	4	9	6	3	8	2	7
7	3	6	4	1	2	5	9
8	9	4	5	6	3	1	2
2	1	5	8	7	9	4	3
9	8	3	1	4	6	7	5
5	2	1	3	8	7	9	6
4	6	7	9	2	5	8	1

INTERMEDIATE - 127

7	8	9	2	3	5	1	4	6
3	5	1	4	7	6	8	9	2
2	4	6	8	1	9	7	3	5
6	7	8	5	4	3	2	1	9
9	2	5	6	8	1	3	7	4
4	1	3	7	9	2	5	6	8
8	3	7	9	5	4	6	2	1
1	9	2	3	6	8	4	5	7
5	6	4	1	2	7	9	8	3

INTERMEDIATE - 128

3	4	2	9	1	6	8	5	7
9	1	5	8	7	4	6	2	3
8	6	7	2	5	3	1	4	9
5	8	6	1	3	7	2	9	4
4	7	9	6	8	2	3	1	5
2	3	1	5	4	9	7	8	6
1	5	4	3	6	8	9	7	2
6	9	8	7	2	5	4	3	1
7	2	3	4	9	1	5	6	8

INTERMEDIATE - 129

4	7	8	2	5	1	3	6	9
9	2	3	6	8	7	1	4	5
1	5	6	9	4	3	8	2	7
6	1	2	7	9	4	5	3	8
7	8	5	1	3	2	6	9	4
3	9	4	8	6	5	2	7	1
5	3	7	4	2	8	9	1	6
8	4	9	3	1	6	7	5	2
2	6	1	5	7	9	4	8	3

INTERMEDIATE - 130

3	7	1	8	6	9	4	5	2
4	5	8	7	2	3	9	6	1
6	9	2	1	5	4	8	3	7
5	2	6	9	3	7	1	4	8
1	8	7	6	4	2	5	9	3
9	4	3	5	1	8	7	2	6
2	3	5	4	8	1	6	7	9
8	6	9	3	7	5	2	1	4
7	1	4	2	9	6	3	8	5

INTERMEDIATE - 131

5	4	7	8	1	2	3	6	9
2	9	1	3	7	6	4	8	5
6	3	8	5	4	9	7	1	2
9	5	2	4	8	3	6	7	1
7	6	4	1	9	5	8	2	3
1	8	3	6	2	7	5	9	4
4	1	6	9	5	8	2	3	7
3	2	9	7	6	4	1	5	8
8	7	5	2	3	1	9	4	6

INTERMEDIATE - 132

5	6	4	7	3	1	8	2
3	2	1	4	9	8	6	5
9	8	7	5	2	6	4	3
7	5	2	3	6	9	1	4
8	4	6	2	1	5	7	9
1	3	9	8	4	7	2	6
6	9	3	1	8	4	5	7
4	7	8	9	5	2	3	1
2	1	5	6	7	3	9	8

INTERMEDIATE - 133

3	1	9	5	6	8	7	4	2
5	8	7	2	4	1	9	6	3
2	6	4	7	3	9	5	1	8
8	4	2	6	9	5	1	3	7
7	3	6	8	1	2	4	5	9
1	9	5	4	7	3	2	8	6
6	7	8	1	2	4	3	9	5
9	2	1	3	5	6	8	7	4
4	5	3	9	8	7	6	2	1

INTERMEDIATE - 134

2	3	4	1	8	9	7	6	5
5	9	7	6	2	3	1	8	4
6	1	8	5	7	4	9	3	2
4	5	3	9	6	1	2	7	8
7	6	1	2	4	8	3	5	9
9	8	2	7	3	5	6	4	1
1	7	9	4	5	6	8	2	3
8	4	6	3	9	2	5	1	7
3	2	5	8	1	7	4	9	6

INTERMEDIATE - 135

5	9	6	8	1	2	4	3	7
3	2	1	7	4	6	9	5	8
7	8	4	5	3	9	2	6	1
8	7	3	2	9	1	5	4	6
4	6	2	3	8	5	1	7	9
9	1	5	6	7	4	8	2	3
2	5	8	9	6	3	7	1	4
6	4	9	1	5	7	3	8	2
1	3	7	4	2	8	6	9	5

INTERMEDIATE - 136

3	6	1	5	2	8	7	4	9
8	9	2	6	7	4	5	1	3
4	7	5	9	3	1	2	6	8
2	1	6	3	5	7	8	9	4
7	8	4	1	9	6	3	5	2
5	3	9	8	4	2	1	7	6
9	4	7	2	8	5	6	3	1
1	5	8	4	6	3	9	2	7
6	2	3	7	1	9	4	8	5

INTERMEDIATE - 137

5	8	4	6	9	2	3	7	1
6	3	7	4	8	1	9	2	5
9	1	2	5	7	3	6	4	8
7	2	6	9	4	5	8	1	3
1	9	5	3	2	8	7	6	4
8	4	3	1	6	7	5	9	2
4	7	1	8	5	9	2	3	6
3	5	9	2	1	6	4	8	7
2	6	8	7	3	4	1	5	9

INTERMEDIATE - 138

5	8	4	1	9	2	7	3
3	9	7	4	5	6	8	2
1	6	2	3	8	7	9	5
7	2	5	9	1	4	6	8
6	1	3	5	7	8	4	9
9	4	8	2	6	3	1	7
4	5	9	7	3	1	2	6
2	7	6	8	4	5	3	1
8	3	1	6	2	9	5	4

INTERMEDIATE - 139

4	3	9	7	5	2	8	6	1
1	6	7	9	3	8	2	5	4
5	8	2	6	4	1	9	7	3
7	1	6	2	9	4	3	8	5
9	2	3	8	6	5	4	1	7
8	4	5	1	7	3	6	2	9
3	7	4	5	2	6	1	9	8
2	9	1	4	8	7	5	3	6
6	5	8	3	1	9	7	4	2

INTERMEDIATE - 140

8	5	2	1	7	3	6	4	9
3	1	9	6	4	5	7	8	2
7	4	6	2	9	8	3	5	1
1	6	7	9	8	4	2	3	5
4	3	5	7	2	6	1	9	8
9	2	8	3	5	1	4	7	6
6	8	1	5	3	7	9	2	4
5	9	3	4	1	2	8	6	7
2	7	4	8	6	9	5	1	3

INTERMEDIATE - 141

4	7	8	6	2	5	1	3	9
5	6	2	9	1	3	4	7	8
9	1	3	4	8	7	5	2	6
6	8	1	5	3	4	7	9	2
2	5	7	8	9	6	3	4	1
3	9	4	2	7	1	8	6	5
1	4	9	3	6	8	2	5	7
7	2	5	1	4	9	6	8	3
8	3	6	7	5	2	9	1	4

INTERMEDIATE - 142

3	1	9	6	5	4	8	2	7
7	5	6	8	9	2	4	1	3
2	8	4	1	7	3	5	6	9
8	9	2	7	4	6	3	5	1
6	7	3	5	8	1	9	4	2
5	4	1	2	3	9	6	7	8
4	3	7	9	1	5	2	8	6
1	2	5	3	6	8	7	9	4
9	6	8	4	2	7	1	3	5

INTERMEDIATE - 143

6	3	5	2	7	4	8	1	9
4	8	1	9	5	6	2	7	3
2	9	7	1	8	3	4	6	5
8	4	6	3	2	5	1	9	7
7	1	3	4	9	8	5	2	6
9	5	2	6	1	7	3	4	8
3	2	8	7	4	9	6	5	1
5	7	4	8	6	1	9	3	2
1	6	9	5	3	2	7	8	4

INTERMEDIATE - 144

5	6	8	2	9	3	1	4
1	3	9	4	5	7	8	2
4	7	2	1	6	8	5	3
8	1	3	9	7	2	4	6
6	9	7	5	8	4	3	1
2	4	5	3	1	6	7	9
3	5	4	8	2	9	6	7
9	8	6	7	3	1	2	5
7	2	1	6	4	5	9	8

INTERMEDIATE - 145

4	9	1	7	8	3	6	5	2
6	2	7	5	9	4	1	8	3
5	3	8	6	2	1	7	4	9
9	1	5	8	3	7	2	6	4
3	4	6	9	5	2	8	7	1
8	7	2	1	4	6	3	9	5
1	8	3	4	6	9	5	2	7
7	5	9	2	1	8	4	3	6
2	6	4	3	7	5	9	1	8

INTERMEDIATE - 146

6	9	4	2	7	1	5	3	8
3	2	7	5	6	8	9	4	1
5	8	1	4	9	3	2	6	7
2	4	3	7	5	9	1	8	6
1	5	9	6	8	4	7	2	3
8	7	6	3	1	2	4	9	5
4	1	2	8	3	5	6	7	9
7	3	5	9	4	6	8	1	2
9	6	8	1	2	7	3	5	4

INTERMEDIATE - 147

6	3	5	8	2	1	4	9	7
4	1	2	3	7	9	5	8	6
7	9	8	4	5	6	1	3	2
8	5	3	1	9	7	6	2	4
9	6	7	2	4	8	3	5	1
1	2	4	6	3	5	9	7	8
3	7	6	5	8	4	2	1	9
5	4	9	7	1	2	8	6	3
2	8	1	9	6	3	7	4	5

INTERMEDIATE - 148

7	4	1	2	9	3	5	6	8
5	9	2	8	7	6	1	3	4
6	3	8	5	4	1	2	7	9
4	8	6	3	1	5	7	9	2
1	2	7	9	6	8	3	4	5
9	5	3	4	2	7	6	8	1
8	7	9	1	3	2	4	5	6
3	1	5	6	8	4	9	2	7
2	6	4	7	5	9	8	1	3

INTERMEDIATE - 149

8	1	2	6	5	3	7	4	9
7	5	9	1	8	4	2	6	3
3	4	6	7	2	9	1	5	8
5	9	1	8	6	7	4	3	2
2	6	8	4	3	1	5	9	7
4	7	3	5	9	2	8	1	6
6	2	4	3	7	5	9	8	1
9	8	5	2	1	6	3	7	4
1	3	7	9	4	8	6	2	5

INTERMEDIATE - 1

3	5	9	8	1	7	2	6	[illegible]
8	4	7	2	6	5	1	3	[illegible]
2	6	1	3	4	9	5	8	7
4	7	3	6	2	1	8	9	5
9	2	5	7	8	3	6	4	1
6	1	8	5	9	4	3	7	2
7	9	6	1	5	8	4	2	3
5	3	2	4	7	6	9	1	8
1	8	4	9	3	2	7	5	6

INTERMEDIATE - 151

4	8	5	1	2	6	3	7	9
1	6	9	5	7	3	4	8	2
7	2	3	4	9	8	5	6	1
8	4	1	6	3	2	9	5	7
2	5	7	9	1	4	8	3	6
9	3	6	7	8	5	2	1	4
6	9	8	2	5	7	1	4	3
3	1	4	8	6	9	7	2	5
5	7	2	3	4	1	6	9	8

INTERMEDIATE - 152

1	3	9	2	6	7	5	8	4
6	2	8	5	4	9	1	3	7
4	7	5	8	1	3	2	6	9
2	6	4	1	7	5	8	9	3
9	8	1	3	2	6	7	4	5
7	5	3	4	9	8	6	2	1
8	9	2	7	3	1	4	5	6
3	4	7	6	5	2	9	1	8
5	1	6	9	8	4	3	7	2

INTERMEDIATE - 153

2	8	3	5	1	7	6	9	4
9	6	1	4	8	2	5	3	7
7	4	5	9	3	6	2	1	8
6	7	9	8	4	5	3	2	1
4	1	2	6	7	3	8	5	9
3	5	8	2	9	1	4	7	6
8	9	7	3	5	4	1	6	2
1	3	6	7	2	8	9	4	5
5	2	4	1	6	9	7	8	3

INTERMEDIATE - 154

2	7	9	4	8	6	5	3	1
6	5	1	9	2	3	8	7	4
8	3	4	5	7	1	2	6	9
3	6	7	2	1	8	9	4	5
1	4	8	7	9	5	3	2	6
9	2	5	6	3	4	7	1	8
4	9	6	3	5	2	1	8	7
7	8	3	1	6	9	4	5	2
5	1	2	8	4	7	6	9	3

INTERMEDIATE - 155

2	3	6	1	7	5	9	8	4
9	1	5	2	4	8	7	3	6
8	7	4	6	3	9	5	1	2
7	9	3	5	8	2	4	6	1
4	2	1	9	6	7	3	5	8
6	5	8	4	1	3	2	9	7
3	4	2	8	5	6	1	7	9
5	8	9	7	2	1	6	4	3
1	6	7	3	9	4	8	2	5

INTERMEDIATE - 156

9	4	1	3	7	6	5	2	8
6	8	2	4	9	5	1	7	3
7	3	5	2	8	1	9	4	6
4	1	3	7	6	9	2	8	5
5	6	9	8	1	2	4	3	7
8	2	7	5	4	3	6	9	1
3	9	6	1	2	8	7	5	4
1	5	4	9	3	7	8	6	2
2	7	8	6	5	4	3	1	9

INTERMEDIATE - 157

6	8	1	5	3	4	2	9	7
5	9	2	7	8	1	6	3	4
4	7	3	9	6	2	5	8	1
8	1	5	2	7	3	9	4	6
7	4	9	6	1	8	3	2	5
3	2	6	4	9	5	7	1	8
1	6	8	3	2	7	4	5	9
9	3	4	1	5	6	8	7	2
2	5	7	8	4	9	1	6	3

INTERMEDIATE - 158

7	4	6	1	3	9	8	5	2
1	9	3	8	5	2	7	4	6
8	2	5	4	7	6	3	9	1
3	5	1	7	2	4	9	6	8
2	6	8	5	9	1	4	7	3
9	7	4	6	8	3	1	2	5
5	3	9	2	4	8	6	1	7
6	8	2	9	1	7	5	3	4
4	1	7	3	6	5	2	8	9

INTERMEDIATE - 159

3	8	1	5	9	2	4	6	7
2	9	7	6	4	1	3	5	8
5	6	4	3	7	8	2	9	1
1	3	9	4	2	5	8	7	6
7	5	8	1	6	3	9	4	2
6	4	2	7	8	9	1	3	5
9	7	3	2	1	6	5	8	4
8	1	6	9	5	4	7	2	3
4	2	5	8	3	7	6	1	9

INTERMEDIATE - 160

7	2	6	9	5	4	8	1	3
5	4	3	2	8	1	7	6	9
9	8	1	3	6	7	4	5	2
6	9	4	7	1	2	3	8	5
8	7	2	6	3	5	9	4	1
1	3	5	4	9	8	2	7	6
3	5	9	8	4	6	1	2	7
2	6	8	1	7	9	5	3	4
4	1	7	5	2	3	6	9	8

INTERMEDIATE - 161

8	6	4	2	3	5	1	7	9
1	9	7	6	8	4	5	2	3
5	3	2	9	7	1	6	8	4
3	2	9	7	1	8	4	5	6
7	4	5	3	9	6	2	1	8
6	1	8	4	5	2	9	3	7
2	8	3	5	4	9	7	6	1
4	5	1	8	6	7	3	9	2
9	7	6	1	2	3	8	4	5

INTERMEDIATE - 162

2	5	9	6	4	1	8	7	3
7	4	8	3	9	2	5	6	1
1	6	3	5	8	7	2	9	4
5	1	7	4	3	9	6	8	2
6	9	4	1	2	8	7	3	5
8	3	2	7	5	6	1	4	9
3	7	6	9	1	5	4	2	8
4	8	1	2	6	3	9	5	7
9	2	5	8	7	4	3	1	6

INTERMEDIATE - 163

6	2	7	1	9	4	5	3	8
3	9	1	8	5	2	6	7	4
4	8	5	3	6	7	2	9	1
7	5	6	4	1	8	9	2	3
8	1	3	9	2	6	7	4	5
9	4	2	5	7	3	1	8	6
1	6	4	7	3	9	8	5	2
2	3	9	6	8	5	4	1	7
5	7	8	2	4	1	3	6	9

INTERMEDIATE - 164

9	3	5	2	4	7	1	8	6
7	4	8	6	3	1	2	9	5
2	1	6	8	9	5	3	4	7
6	9	2	1	8	4	7	5	3
1	8	7	5	6	3	4	2	9
4	5	3	7	2	9	6	1	8
3	7	9	4	5	2	8	6	1
5	6	4	3	1	8	9	7	2
8	2	1	9	7	6	5	3	4

INTERMEDIATE - 165

6	8	9	3	7	2	5	1	4
7	2	5	6	4	1	8	9	3
1	3	4	5	9	8	7	2	6
8	5	6	4	1	7	9	3	2
2	1	7	9	5	3	6	4	8
4	9	3	8	2	6	1	7	5
3	7	1	2	8	5	4	6	9
9	6	8	7	3	4	2	5	1
5	4	2	1	6	9	3	8	7

INTERMEDIATE - 166

8	3	1	7	4	9	2	6	5
6	5	4	1	2	8	3	9	7
9	2	7	5	6	3	8	4	1
3	4	8	6	9	7	1	5	2
7	1	9	2	8	5	4	3	6
2	6	5	4	3	1	9	7	8
4	8	6	9	5	2	7	1	3
1	9	3	8	7	6	5	2	4
5	7	2	3	1	4	6	8	9

INTERMEDIATE - 167

4	8	3	9	7	1	2	5	6
2	7	5	6	3	8	4	9	1
1	6	9	5	4	2	3	8	7
7	5	1	2	8	4	9	6	3
3	2	6	7	5	9	8	1	4
8	9	4	3	1	6	5	7	2
5	4	8	1	2	7	6	3	9
9	1	2	8	6	3	7	4	5
6	3	7	4	9	5	1	2	8

INTERMEDIATE - 168

5	4	6	9	8	1	2	3	7
2	7	9	4	3	6	5	1	8
8	3	1	5	7	2	4	6	9
4	9	3	7	6	5	8	2	1
1	8	5	2	9	4	3	7	6
7	6	2	3	1	8	9	4	5
3	5	7	6	4	9	1	8	2
6	2	8	1	5	3	7	9	4
9	1	4	8	2	7	6	5	3

INTERMEDIATE - 169

9	1	4	6	3	7	8	5	2
7	3	2	9	8	5	4	1	6
6	5	8	4	2	1	3	7	9
4	2	1	7	6	9	5	3	8
8	9	6	3	5	4	1	2	7
3	7	5	8	1	2	9	6	4
1	6	3	2	4	8	7	9	5
2	4	7	5	9	3	6	8	1
5	8	9	1	7	6	2	4	3

INTERMEDIATE - 170

1	7	8	3	6	5	9	4	2
2	6	5	4	9	8	1	7	3
3	9	4	1	2	7	5	6	8
6	1	7	2	5	4	3	8	9
5	3	2	8	7	9	6	1	4
8	4	9	6	1	3	7	2	5
4	8	1	9	3	6	2	5	7
9	5	6	7	8	2	4	3	1
7	2	3	5	4	1	8	9	6

INTERMEDIATE - 171

2	5	6	9	4	3	7	8	1
9	1	4	7	5	8	2	6	3
7	3	8	1	2	6	4	9	5
3	6	5	4	9	1	8	7	2
1	8	9	2	6	7	3	5	4
4	7	2	8	3	5	9	1	6
5	4	7	3	1	9	6	2	8
6	9	3	5	8	2	1	4	7
8	2	1	6	7	4	5	3	9

INTERMEDIATE - 172

1	3	6	5	8	2	9	4	7
2	5	9	3	4	7	8	6	1
8	4	7	9	1	6	5	2	3
6	8	3	1	9	5	2	7	4
9	7	5	2	6	4	3	1	8
4	2	1	7	3	8	6	9	5
7	1	8	6	2	3	4	5	9
5	6	4	8	7	9	1	3	2
3	9	2	4	5	1	7	8	6

INTERMEDIATE -173

3	9	1	5	8	4	6	2	7
4	7	8	9	6	2	1	5	3
5	6	2	3	7	1	9	4	8
2	4	6	1	9	3	7	8	5
9	8	7	6	2	5	3	1	4
1	3	5	7	4	8	2	6	9
8	2	9	4	3	6	5	7	1
7	5	4	2	1	9	8	3	6
6	1	3	8	5	7	4	9	2

INTERMEDIATE - 174

3	4	7	1	6	8	9	5	2
6	9	2	7	5	4	3	1	8
1	5	8	3	2	9	6	7	4
5	8	3	6	1	2	4	9	7
4	2	6	9	7	3	1	8	5
9	7	1	4	8	5	2	3	6
2	3	9	8	4	7	5	6	1
8	1	5	2	3	6	7	4	9
7	6	4	5	9	1	8	2	3

INTERMEDIATE - 175

9	5	1	8	4	2	3	6	7
6	8	2	1	7	3	9	5	4
7	4	3	9	5	6	8	2	1
5	3	9	4	2	7	1	8	6
2	7	8	6	9	1	5	4	3
1	6	4	5	3	8	2	7	9
8	9	5	3	6	4	7	1	2
4	1	7	2	8	9	6	3	5
3	2	6	7	1	5	4	9	8

INTERMEDIATE - 176

7	6	2	1	8	9	3	4	5
8	1	5	3	4	2	7	6	9
4	9	3	6	7	5	2	1	8
9	3	4	2	5	8	1	7	6
1	2	8	7	6	3	5	9	4
6	5	7	9	1	4	8	3	2
5	8	6	4	3	1	9	2	7
2	4	1	8	9	7	6	5	3
3	7	9	5	2	6	4	8	1

INTERMEDIATE - 177

5	9	2	4	3	7	6	8	1
6	1	4	8	2	5	7	9	3
8	3	7	9	1	6	2	5	4
1	6	3	2	4	8	5	7	9
7	5	9	1	6	3	4	2	8
2	4	8	5	7	9	1	3	6
3	7	5	6	8	1	9	4	2
9	2	6	3	5	4	8	1	7
4	8	1	7	9	2	3	6	5

INTERMEDIATE - 178

5	3	2	8	9	6	1	7	4
6	8	1	4	5	7	2	9	3
4	7	9	1	2	3	5	8	6
2	1	7	9	4	8	6	3	5
9	6	8	5	3	2	4	1	7
3	4	5	7	6	1	8	2	9
1	9	4	3	8	5	7	6	2
8	5	6	2	7	9	3	4	1
7	2	3	6	1	4	9	5	8

INTERMEDIATE - 179

5	1	8	2	9	6	3	7	4
7	4	9	3	1	5	8	2	6
3	2	6	4	8	7	9	1	5
9	3	7	8	5	2	6	4	1
2	6	1	9	3	4	5	8	7
4	8	5	7	6	1	2	3	9
6	5	2	1	7	3	4	9	8
8	7	4	6	2	9	1	5	3
1	9	3	5	4	8	7	6	2

INTERMEDIATE - 180

6	2	3	7	1	8	4	9	5
1	4	5	6	9	3	7	2	8
7	9	8	4	2	5	6	1	3
2	8	4	3	5	6	9	7	1
9	7	6	1	8	2	5	3	4
5	3	1	9	7	4	2	8	6
4	5	2	8	3	9	1	6	7
8	1	9	5	6	7	3	4	2
3	6	7	2	4	1	8	5	9

INTERMEDIATE - 181

3	7	4	2	1	6	9	8	5
1	6	8	9	5	3	2	4	7
2	5	9	4	8	7	3	1	6
6	8	1	3	7	5	4	2	9
4	2	5	1	6	9	7	3	8
7	9	3	8	4	2	5	6	1
5	4	2	6	9	8	1	7	3
8	1	7	5	3	4	6	9	2
9	3	6	7	2	1	8	5	4

INTERMEDIATE - 182

9	5	2	4	6	3	1	7	8
6	8	1	2	9	7	4	3	5
7	4	3	5	1	8	2	9	6
4	1	5	3	8	2	7	6	9
2	6	7	9	5	4	3	8	1
8	3	9	1	7	6	5	4	2
5	2	4	8	3	9	6	1	7
1	9	6	7	4	5	8	2	3
3	7	8	6	2	1	9	5	4

INTERMEDIATE - 183

6	2	1	8	5	4	3	7	9
9	8	5	2	3	7	6	1	4
7	3	4	1	9	6	8	2	5
5	1	9	6	4	8	7	3	2
2	4	8	3	7	9	1	5	6
3	6	7	5	2	1	4	9	8
4	7	2	9	8	3	5	6	1
1	9	3	4	6	5	2	8	7
8	5	6	7	1	2	9	4	3

INTERMEDIATE - 184

3	2	5	6	1	4	8	7	9
7	6	1	5	9	8	4	2	3
4	9	8	3	7	2	6	5	1
5	8	6	7	2	1	3	9	4
1	7	3	4	5	9	2	6	8
2	4	9	8	6	3	5	1	7
9	3	2	1	8	5	7	4	6
6	5	4	9	3	7	1	8	2
8	1	7	2	4	6	9	3	5

INTERMEDIATE - 185

9	4	2	3	8	1	6	7	5
7	1	6	5	4	2	3	8	9
8	3	5	6	7	9	2	1	4
3	6	1	7	2	4	5	9	8
4	8	7	9	5	6	1	3	2
5	2	9	1	3	8	7	4	6
6	5	4	8	1	3	9	2	7
1	9	8	2	6	7	4	5	3
2	7	3	4	9	5	8	6	1

INTERMEDIATE - 186

7	2	9	8	1	3	4	6	5
5	6	8	9	2	4	1	7	3
1	3	4	7	5	6	8	2	9
4	9	6	5	8	2	3	1	7
8	7	5	6	3	1	9	4	2
2	1	3	4	9	7	5	8	6
9	4	1	2	7	5	6	3	8
3	8	7	1	6	9	2	5	4
6	5	2	3	4	8	7	9	1

INTERMEDIATE - 187

3	1	8	6	4	5	2	7	9
6	2	9	8	7	1	5	3	4
5	7	4	2	3	9	8	6	1
2	4	3	1	8	7	9	5	6
9	8	5	3	6	4	1	2	7
7	6	1	9	5	2	4	8	3
8	9	7	5	1	6	3	4	2
4	5	2	7	9	3	6	1	8
1	3	6	4	2	8	7	9	5

INTERMEDIATE - 188

4	5	3	9	1	6	8	7	2
2	8	7	3	5	4	9	6	1
1	6	9	2	8	7	3	4	5
7	4	2	5	6	9	1	8	3
8	9	5	7	3	1	4	2	6
6	3	1	4	2	8	7	5	9
5	1	6	8	4	3	2	9	7
3	7	8	6	9	2	5	1	4
9	2	4	1	7	5	6	3	8

INTERMEDIATE - 189

3	2	6	9	7	8	5	1	4
8	1	9	6	4	5	2	3	7
5	7	4	1	2	3	6	8	9
1	4	5	7	6	9	8	2	3
9	3	7	8	5	2	4	6	1
6	8	2	3	1	4	7	9	5
4	6	8	5	3	1	9	7	2
7	5	1	2	9	6	3	4	8
2	9	3	4	8	7	1	5	6

INTERMEDIATE - 190

4	2	5	3	7	6	1	9	8
7	8	3	1	2	9	4	6	5
6	9	1	5	8	4	7	2	3
2	1	6	9	3	5	8	7	4
8	7	4	2	6	1	3	5	9
3	5	9	7	4	8	6	1	2
9	3	2	8	1	7	5	4	6
5	4	7	6	9	3	2	8	1
1	6	8	4	5	2	9	3	7

INTERMEDIATE - 191

1	5	2	8	6	4	9	3	7
4	6	3	7	9	1	2	8	5
9	7	8	3	5	2	1	4	6
3	1	4	5	7	9	8	6	2
2	9	7	4	8	6	3	5	1
6	8	5	1	2	3	4	7	9
7	3	1	2	4	5	6	9	8
8	2	6	9	3	7	5	1	4
5	4	9	6	1	8	7	2	3

INTERMEDIATE - 192

1	9	3	4	6	2	8	7	5
6	2	5	7	8	1	4	3	9
4	8	7	9	3	5	1	2	6
9	6	2	1	5	4	7	8	3
5	7	1	8	9	3	2	6	4
3	4	8	6	2	7	9	5	1
7	5	9	3	1	8	6	4	2
2	1	4	5	7	6	3	9	8
8	3	6	2	4	9	5	1	7

INTERMEDIATE - 193

2	7	1	8	5	6	9	4	3
8	4	9	7	1	3	2	6	5
6	3	5	2	4	9	7	8	1
7	5	8	9	6	2	3	1	4
9	1	3	4	7	5	8	2	6
4	6	2	1	3	8	5	9	7
3	2	6	5	9	1	4	7	8
5	9	7	6	8	4	1	3	2
1	8	4	3	2	7	6	5	9

INTERMEDIATE - 194

7	6	3	2	1	4	5	9	8
9	4	1	8	5	7	2	6	3
5	8	2	3	9	6	4	1	7
4	2	9	7	3	1	8	5	6
6	3	8	5	2	9	7	4	1
1	7	5	4	6	8	9	3	2
2	5	6	9	8	3	1	7	4
8	1	7	6	4	5	3	2	9
3	9	4	1	7	2	6	8	5

INTERMEDIATE - 195

3	4	8	7	2	5	1	9	6
6	5	1	4	9	3	8	7	2
9	2	7	1	6	8	3	5	4
7	6	2	9	8	1	4	3	5
8	1	9	3	5	4	6	2	7
4	3	5	2	7	6	9	8	1
5	9	3	6	1	7	2	4	8
1	8	4	5	3	2	7	6	9
2	7	6	8	4	9	5	1	3

INTERMEDIATE - 196

9	3	7	6	2	4	8	1	5
2	1	6	8	5	7	9	4	3
4	8	5	3	9	1	7	6	2
7	6	9	1	3	2	4	5	8
8	2	3	5	4	6	1	9	7
1	5	4	7	8	9	3	2	6
6	7	2	9	1	3	5	8	4
3	9	8	4	6	5	2	7	1
5	4	1	2	7	8	6	3	9

INTERMEDIATE - 197

8	5	6	2	7	4	1	3	9
9	1	7	3	5	8	6	4	2
4	3	2	6	9	1	7	8	5
7	6	8	4	1	9	2	5	3
3	4	9	7	2	5	8	1	6
5	2	1	8	3	6	4	9	7
1	8	3	9	6	7	5	2	4
2	7	4	5	8	3	9	6	1
6	9	5	1	4	2	3	7	8

INTERMEDIATE - 198

9	7	3	2	5	8	1	4	6
4	8	5	6	1	3	2	7	9
6	1	2	7	4	9	8	3	5
8	3	4	1	6	7	5	9	2
5	9	6	4	3	2	7	1	8
1	2	7	9	8	5	3	6	4
2	6	9	5	7	1	4	8	3
7	5	8	3	9	4	6	2	1
3	4	1	8	2	6	9	5	7

INTERMEDIATE - 199

6	8	5	2	4	1	3	9	7
4	9	2	7	6	3	1	8	5
7	3	1	8	5	9	2	6	4
8	1	6	3	9	5	4	7	2
9	2	3	4	7	6	5	1	8
5	4	7	1	8	2	6	3	9
1	5	8	9	3	4	7	2	6
2	7	4	6	1	8	9	5	3
3	6	9	5	2	7	8	4	1

INTERMEDIATE - 200

3	1	6	4	5	8	9	7	2
2	5	4	9	6	7	1	8	3
7	9	8	3	1	2	4	6	5
5	6	7	2	8	1	3	4	9
4	8	3	6	9	5	2	1	7
1	2	9	7	3	4	6	5	8
8	3	2	5	4	6	7	9	1
6	7	1	8	2	9	5	3	4
9	4	5	1	7	3	8	2	6

INTERMEDIATE - 201

3	4	9	7	6	2	1	8	5
7	6	8	1	5	9	3	4	2
1	5	2	4	3	8	6	7	9
9	2	7	5	4	6	8	1	3
4	3	1	2	8	7	9	5	6
6	8	5	3	9	1	7	2	4
2	9	3	8	7	4	5	6	1
5	7	4	6	1	3	2	9	8
8	1	6	9	2	5	4	3	7

INTERMEDIATE - 202

9	4	5	1	6	3	8	7	2
8	3	1	2	7	5	9	4	6
2	7	6	8	4	9	1	3	5
5	2	8	4	3	1	6	9	7
1	9	4	7	8	6	5	2	3
3	6	7	9	5	2	4	1	8
6	1	2	3	9	8	7	5	4
7	8	3	5	1	4	2	6	9
4	5	9	6	2	7	3	8	1

INTERMEDIATE - 203

5	7	9	6	1	2	3	4	8
1	2	4	8	3	9	5	6	7
3	6	8	5	7	4	9	1	2
4	3	7	2	9	6	1	8	5
8	1	2	7	5	3	6	9	4
6	9	5	4	8	1	7	2	3
7	4	3	9	6	8	2	5	1
2	5	6	1	4	7	8	3	9
9	8	1	3	2	5	4	7	6

INTERMEDIATE - 204

6	3	9	8	2	4	5	1	7
2	7	4	5	1	9	3	6	8
5	8	1	7	6	3	9	2	4
4	6	7	1	5	2	8	9	3
9	2	5	6	3	8	4	7	1
3	1	8	9	4	7	6	5	2
8	9	3	2	7	5	1	4	6
1	5	2	4	8	6	7	3	9
7	4	6	3	9	1	2	8	5

INTERMEDIATE - 205

3	9	6	5	8	2	7	4	1
4	1	2	3	7	9	5	8	6
8	7	5	6	4	1	2	3	9
1	6	9	8	3	7	4	2	5
2	8	4	1	6	5	9	7	3
7	5	3	2	9	4	1	6	8
6	4	1	7	5	3	8	9	2
5	3	7	9	2	8	6	1	4
9	2	8	4	1	6	3	5	7

INTERMEDIATE - 206

3	9	8	5	2	6	7	4	1
7	2	1	9	3	4	6	8	5
6	5	4	7	8	1	9	2	3
4	3	6	2	1	7	5	9	8
9	1	2	8	6	5	4	3	7
8	7	5	3	4	9	1	6	2
5	8	7	6	9	3	2	1	4
1	6	3	4	7	2	8	5	9
2	4	9	1	5	8	3	7	6

INTERMEDIATE - 207

5	9	4	2	1	6	8	7	3
8	2	1	7	4	3	9	5	6
6	7	3	9	8	5	1	4	2
9	8	2	6	7	1	5	3	4
3	1	7	4	5	9	6	2	8
4	6	5	8	3	2	7	9	1
1	3	6	5	9	4	2	8	7
2	5	8	3	6	7	4	1	9
7	4	9	1	2	8	3	6	5

INTERMEDIATE - 208

1	7	5	4	9	3	2	8	6
3	8	4	6	5	2	1	9	7
2	9	6	8	1	7	3	4	5
6	1	3	2	4	5	8	7	9
7	2	9	3	8	6	5	1	4
5	4	8	9	7	1	6	3	2
8	6	7	1	2	9	4	5	3
9	3	1	5	6	4	7	2	8
4	5	2	7	3	8	9	6	1

INTERMEDIATE - 209

6	4	1	2	8	9	3	5	7
9	7	2	5	3	6	1	4	8
8	3	5	4	1	7	9	2	6
3	1	8	7	6	4	5	9	2
7	5	9	3	2	1	8	6	4
4	2	6	8	9	5	7	1	3
5	9	7	6	4	8	2	3	1
1	6	3	9	7	2	4	8	5
2	8	4	1	5	3	6	7	9

INTERMEDIATE - 210

6	9	7	1	3	2	8	4	5
8	2	1	4	5	9	6	7	3
5	4	3	8	7	6	2	1	9
7	5	9	3	6	4	1	2	8
4	1	8	9	2	7	5	3	6
3	6	2	5	8	1	7	9	4
9	8	4	7	1	5	3	6	2
2	7	5	6	9	3	4	8	1
1	3	6	2	4	8	9	5	7

INTERMEDIATE - 211

9	8	6	2	7	4	1	5	3
7	2	4	1	3	5	9	8	6
5	3	1	6	8	9	2	4	7
4	9	7	3	5	2	6	1	8
3	6	2	8	4	1	7	9	5
8	1	5	9	6	7	4	3	2
2	4	8	7	1	3	5	6	9
6	5	9	4	2	8	3	7	1
1	7	3	5	9	6	8	2	4

INTERMEDIATE - 212

4	5	3	1	7	2	6	9	8
6	2	8	9	5	4	7	1	3
1	7	9	8	6	3	4	2	5
3	1	4	5	2	6	8	7	9
8	6	7	3	1	9	5	4	2
2	9	5	7	4	8	3	6	1
9	8	6	2	3	7	1	5	4
7	3	1	4	9	5	2	8	6
5	4	2	6	8	1	9	3	7

INTERMEDIATE - 213

5	6	9	8	1	4	7	3	2
4	1	8	3	2	7	5	9	6
7	3	2	5	9	6	4	8	1
6	8	7	1	3	9	2	5	4
9	4	5	2	7	8	1	6	3
1	2	3	6	4	5	8	7	9
2	7	6	9	8	1	3	4	5
8	9	1	4	5	3	6	2	7
3	5	4	7	6	2	9	1	8

INTERMEDIATE - 214

6	8	3	5	9	2	7	1	4
7	1	4	6	8	3	2	9	5
5	9	2	4	7	1	8	6	3
1	7	8	2	3	6	5	4	9
3	4	9	8	5	7	1	2	6
2	5	6	1	4	9	3	7	8
4	6	5	7	1	8	9	3	2
8	3	1	9	2	4	6	5	7
9	2	7	3	6	5	4	8	1

INTERMEDIATE - 215

5	4	1	9	3	7	8	6	2
8	9	3	1	6	2	4	5	7
7	2	6	8	5	4	3	1	9
4	6	8	3	2	9	5	7	1
3	1	9	7	8	5	6	2	4
2	5	7	4	1	6	9	3	8
9	7	5	2	4	3	1	8	6
1	3	4	6	7	8	2	9	5
6	8	2	5	9	1	7	4	3

INTERMEDIATE - 216

4	6	3	1	2	5	7	8	9
8	9	7	3	6	4	2	5	1
1	2	5	8	9	7	3	4	6
2	4	1	5	7	6	9	3	8
3	8	6	4	1	9	5	7	2
5	7	9	2	3	8	6	1	4
7	3	2	6	4	1	8	9	5
6	1	8	9	5	3	4	2	7
9	5	4	7	8	2	1	6	3

INTERMEDIATE - 217

3	7	9	1	6	2	8	5	4
5	1	4	8	7	9	6	2	3
6	2	8	5	3	4	1	9	7
1	3	2	9	4	7	5	6	8
4	6	7	2	8	5	9	3	1
9	8	5	6	1	3	4	7	2
7	9	1	3	5	8	2	4	6
2	4	6	7	9	1	3	8	5
8	5	3	4	2	6	7	1	9

INTERMEDIATE - 218

6	7	9	5	2	8	1	3	4
3	4	5	9	1	6	2	8	7
8	2	1	7	3	4	9	6	5
9	3	7	1	4	5	8	2	6
1	8	2	6	9	7	5	4	3
5	6	4	3	8	2	7	9	1
4	5	3	2	7	9	6	1	8
7	9	8	4	6	1	3	5	2
2	1	6	8	5	3	4	7	9

INTERMEDIATE - 219

8	7	2	4	6	5	9	3	1
3	5	9	8	2	1	6	4	7
6	4	1	9	3	7	8	5	2
5	6	3	1	7	9	2	8	4
4	1	8	2	5	3	7	6	9
9	2	7	6	8	4	5	1	3
7	9	6	3	4	8	1	2	5
2	3	5	7	1	6	4	9	8
1	8	4	5	9	2	3	7	6

INTERMEDIATE - 220

4	3	2	9	1	7	6	5	8
9	7	5	6	8	4	1	3	2
8	1	6	5	3	2	9	7	4
5	6	8	7	2	9	4	1	3
7	9	1	8	4	3	5	2	6
2	4	3	1	5	6	8	9	7
1	8	4	2	7	5	3	6	9
3	2	9	4	6	1	7	8	5
6	5	7	3	9	8	2	4	1

INTERMEDIATE - 221

4	8	3	1	9	7	5	6	2
2	5	1	6	4	8	7	3	9
7	9	6	5	3	2	8	4	1
3	4	7	8	1	6	2	9	5
5	1	9	3	2	4	6	8	7
8	6	2	9	7	5	3	1	4
9	7	4	2	8	3	1	5	6
1	3	5	7	6	9	4	2	8
6	2	8	4	5	1	9	7	3

INTERMEDIATE - 222

5	4	3	2	9	6	8	1	[illegible]
8	6	2	7	5	1	9	3	[illegible]
9	7	1	4	8	3	2	6	[illegible]
2	9	4	6	1	5	3	7	[illegible]
1	3	6	8	7	2	5	4	[illegible]
7	8	5	3	4	9	6	2	[illegible]
6	5	9	1	2	4	7	8	[illegible]
3	1	8	9	6	7	4	5	[illegible]
4	2	7	5	3	8	1	9	[illegible]

INTERMEDIATE - 223

6	5	2	9	3	8	4	1	7
7	9	1	2	4	5	6	8	3
8	3	4	6	7	1	2	5	9
3	8	9	1	5	6	7	4	2
5	4	6	7	8	2	9	3	1
2	1	7	4	9	3	5	6	8
1	7	5	8	6	9	3	2	4
9	6	8	3	2	4	1	7	5
4	2	3	5	1	7	8	9	6

INTERMEDIATE - 224

7	8	5	2	6	3	4	1	9
6	2	3	9	4	1	5	7	8
1	4	9	8	7	5	3	2	6
9	6	2	4	3	8	7	5	1
8	1	7	6	5	2	9	3	4
3	5	4	7	1	9	6	8	2
2	7	6	5	8	4	1	9	3
5	9	1	3	2	6	8	4	7
4	3	8	1	9	7	2	6	5

INTERMEDIATE - 225

6	3	9	2	4	1	8	7	5
1	2	7	5	3	8	6	4	9
8	5	4	9	7	6	3	1	2
7	1	8	6	2	9	5	3	4
3	6	5	1	8	4	2	9	7
4	9	2	7	5	3	1	8	6
2	4	1	3	9	5	7	6	8
9	7	3	8	6	2	4	5	1
5	8	6	4	1	7	9	2	3

INTERMEDIATE - 226

7	4	5	3	6	8	2	9	1
1	8	9	7	2	5	3	6	4
6	3	2	9	1	4	5	8	7
2	5	1	6	8	7	9	4	3
8	7	3	2	4	9	1	5	6
4	9	6	5	3	1	7	2	8
3	6	7	4	9	2	8	1	5
9	1	4	8	5	3	6	7	2
5	2	8	1	7	6	4	3	9

INTERMEDIATE - 227

1	2	8	4	7	5	6	9	3
3	9	7	1	6	8	4	2	5
5	6	4	2	9	3	8	7	1
2	8	3	9	1	4	5	6	7
7	5	9	8	3	6	1	4	2
6	4	1	7	5	2	3	8	9
4	3	2	5	8	7	9	1	6
9	7	6	3	4	1	2	5	8
8	1	5	6	2	9	7	3	4

INTERMEDIATE - 228

6	4	8	1	5	2	7	3
1	3	5	7	9	4	8	2
9	2	7	3	8	6	4	1
5	1	6	9	2	8	3	7
3	9	2	4	6	7	1	5
8	7	4	5	1	3	6	9
7	8	9	6	3	5	2	4
4	6	1	2	7	9	5	8
2	5	3	8	4	1	9	6

INTERMEDIATE - 229

7	9	1	3	5	6	8	2	4
8	4	3	2	9	1	5	7	6
6	2	5	4	7	8	9	3	1
1	3	9	5	4	2	7	6	8
4	6	8	1	3	7	2	5	9
2	5	7	8	6	9	1	4	3
9	8	4	6	2	5	3	1	7
3	7	2	9	1	4	6	8	5
5	1	6	7	8	3	4	9	2

INTERMEDIATE - 230

8	4	2	3	9	7	1	5	6
7	6	9	5	1	4	8	2	3
5	1	3	6	2	8	4	7	9
1	8	5	7	6	9	2	3	4
6	2	4	8	3	1	7	9	5
3	9	7	4	5	2	6	8	1
4	3	8	1	7	5	9	6	2
9	7	6	2	4	3	5	1	8
2	5	1	9	8	6	3	4	7

INTERMEDIATE - 231

2	8	1	3	4	5	9	7	6
5	6	7	9	2	8	4	3	1
9	4	3	7	1	6	5	2	8
7	3	8	4	5	9	6	1	2
1	2	4	8	6	7	3	9	5
6	9	5	2	3	1	8	4	7
3	5	6	1	7	4	2	8	9
4	1	9	6	8	2	7	5	3
8	7	2	5	9	3	1	6	4

INTERMEDIATE - 232

6	4	7	5	8	3	2	1	9
2	9	8	4	7	1	5	6	3
3	1	5	9	6	2	4	7	8
1	2	4	8	5	9	7	3	6
7	3	6	1	2	4	9	8	5
8	5	9	7	3	6	1	2	4
4	8	2	6	1	5	3	9	7
9	6	1	3	4	7	8	5	2
5	7	3	2	9	8	6	4	1

INTERMEDIATE - 233

5	4	9	1	8	2	3	7	6
7	6	8	3	5	4	2	9	1
3	1	2	6	7	9	4	5	8
8	5	1	7	6	3	9	4	2
4	2	7	8	9	1	6	3	5
6	9	3	2	4	5	8	1	7
2	7	4	9	1	8	5	6	3
9	8	6	5	3	7	1	2	4
1	3	5	4	2	6	7	8	9

INTERMEDIATE - 234

9	4	8	6	7	3	2	5
7	3	1	2	5	9	6	4
6	5	2	8	4	1	3	9
4	2	7	9	8	5	1	3
5	1	6	3	2	7	4	8
3	8	9	1	6	4	7	2
8	9	3	4	1	6	5	7
1	7	4	5	9	2	8	6
2	6	5	7	3	8	9	1

INTERMEDIATE - 235

2	9	1	3	4	6	7	5	8
5	4	8	1	9	7	6	3	2
6	3	7	8	5	2	1	9	4
4	1	2	6	7	9	5	8	3
9	7	5	2	3	8	4	6	1
8	6	3	4	1	5	9	2	7
3	8	9	7	6	4	2	1	5
7	2	6	5	8	1	3	4	9
1	5	4	9	2	3	8	7	6

INTERMEDIATE - 236

3	2	5	4	8	9	1	6	7
9	6	1	7	5	2	4	8	3
4	8	7	6	1	3	2	9	5
5	1	4	2	7	6	8	3	9
8	3	9	1	4	5	7	2	6
6	7	2	9	3	8	5	1	4
7	4	3	8	9	1	6	5	2
2	5	8	3	6	7	9	4	1
1	9	6	5	2	4	3	7	8

INTERMEDIATE - 237

6	8	7	3	9	4	5	1	2
2	1	5	8	6	7	9	3	4
4	3	9	5	2	1	6	7	8
5	4	8	6	1	3	7	2	9
9	6	3	2	7	8	1	4	5
7	2	1	9	4	5	8	6	3
8	5	4	7	3	6	2	9	1
3	7	2	1	8	9	4	5	6
1	9	6	4	5	2	3	8	7

INTERMEDIATE - 238

5	3	8	2	9	6	7	4	1
6	9	4	7	1	3	2	8	5
7	1	2	4	8	5	9	6	3
9	6	3	5	4	8	1	7	2
4	2	1	6	7	9	3	5	8
8	5	7	3	2	1	6	9	4
1	8	5	9	3	7	4	2	6
2	7	6	1	5	4	8	3	9
3	4	9	8	6	2	5	1	7

INTERMEDIATE - 239

6	3	7	4	1	9	8	5	2
8	1	2	5	6	7	4	3	9
4	5	9	2	8	3	6	1	7
2	7	1	9	5	4	3	8	6
5	9	8	3	2	6	7	4	1
3	6	4	1	7	8	9	2	5
1	4	6	8	9	5	2	7	3
7	2	3	6	4	1	5	9	8
9	8	5	7	3	2	1	6	4

INTERMEDIATE - 240

6	3	2	9	7	4	8	1
1	7	9	3	5	8	2	4
5	4	8	6	2	1	7	9
8	5	3	4	9	6	1	2
9	2	6	1	3	7	5	8
4	1	7	5	8	2	3	6
7	8	5	2	6	9	4	3
2	6	4	7	1	3	9	5
3	9	1	8	4	5	6	7

INTERMEDIATE - 241

4	7	3	9	6	2	5	1	8
1	2	9	5	7	8	4	6	3
5	8	6	4	1	3	2	7	9
3	9	2	6	5	1	8	4	7
7	1	8	3	2	4	9	5	6
6	4	5	7	8	9	3	2	1
8	6	4	1	9	5	7	3	2
2	3	1	8	4	7	6	9	5
9	5	7	2	3	6	1	8	4

INTERMEDIATE - 242

5	8	2	1	6	7	4	9	3
4	6	9	5	2	3	1	7	8
7	3	1	9	8	4	6	2	5
8	1	4	7	3	9	5	6	2
3	7	5	2	4	6	9	8	1
9	2	6	8	5	1	7	3	4
1	5	3	6	9	2	8	4	7
2	9	8	4	7	5	3	1	6
6	4	7	3	1	8	2	5	9

INTERMEDIATE - 243

8	9	6	3	4	5	1	7	2
4	5	7	1	2	9	6	3	8
2	3	1	8	6	7	4	9	5
9	1	2	6	7	4	8	5	3
6	8	4	5	3	2	9	1	7
5	7	3	9	8	1	2	6	4
3	6	9	4	5	8	7	2	1
1	2	8	7	9	3	5	4	6
7	4	5	2	1	6	3	8	9

INTERMEDIATE - 244

2	7	6	3	8	5	9	4	1
4	8	3	1	2	9	6	5	7
5	1	9	6	4	7	2	8	3
1	2	8	4	7	6	3	9	5
9	3	4	5	1	2	7	6	8
7	6	5	9	3	8	1	2	4
8	4	2	7	6	3	5	1	9
3	5	1	2	9	4	8	7	6
6	9	7	8	5	1	4	3	2

INTERMEDIATE - 245

2	9	8	3	7	5	4	6	1
1	6	5	8	4	2	3	7	9
4	3	7	9	1	6	8	2	5
3	1	6	5	8	9	7	4	2
7	2	9	4	3	1	5	8	6
8	5	4	6	2	7	9	1	3
6	8	2	7	9	3	1	5	4
9	7	1	2	5	4	6	3	8
5	4	3	1	6	8	2	9	7

INTERMEDIATE - 246

6	2	5	7	9	4	3	8	1
3	9	1	8	5	6	4	7	2
7	8	4	2	3	1	6	9	5
9	1	8	4	7	3	5	2	6
5	3	2	6	1	8	7	4	9
4	7	6	5	2	9	8	1	3
8	5	7	1	6	2	9	3	4
1	4	3	9	8	5	2	6	7
2	6	9	3	4	7	1	5	8

INTERMEDIATE - 247

8	6	9	7	5	3	1	4	2
5	1	3	2	9	4	8	6	7
2	4	7	6	8	1	5	9	3
3	5	6	4	2	9	7	8	1
9	7	1	3	6	8	2	5	4
4	8	2	1	7	5	9	3	6
7	3	8	5	4	2	6	1	9
1	2	5	9	3	6	4	7	8
6	9	4	8	1	7	3	2	5

INTERMEDIATE - 248

9	6	1	5	7	2	8	3	4
3	4	2	8	9	6	7	5	1
7	8	5	1	3	4	2	6	9
5	1	7	4	2	9	6	8	3
8	2	4	6	5	3	9	1	7
6	9	3	7	1	8	5	4	2
4	3	8	9	6	7	1	2	5
1	7	6	2	4	5	3	9	8
2	5	9	3	8	1	4	7	6

INTERMEDIATE - 249

5	4	7	1	6	2	3	8	9
3	6	9	7	4	8	1	5	2
8	1	2	3	5	9	7	6	4
7	8	6	4	2	1	5	9	3
2	5	1	9	8	3	4	7	6
9	3	4	5	7	6	8	2	1
4	9	5	6	3	7	2	1	8
6	7	8	2	1	4	9	3	5
1	2	3	8	9	5	6	4	7

INTERMEDIATE - 250

1	9	3	2	6	8	7	4	5
7	4	6	9	1	5	2	3	8
2	8	5	4	3	7	1	9	6
9	5	7	6	4	2	3	8	1
3	6	8	1	5	9	4	2	7
4	2	1	8	7	3	5	6	9
5	1	4	3	8	6	9	7	2
6	3	2	7	9	1	8	5	4
8	7	9	5	2	4	6	1	3

INTERMEDIATE - 251

8	3	5	4	1	9	2	6	7
6	9	1	7	8	2	5	3	4
2	4	7	6	3	5	9	1	8
9	7	3	2	4	8	6	5	1
1	6	2	5	7	3	8	4	9
4	5	8	1	9	6	3	7	2
3	1	9	8	6	4	7	2	5
7	2	6	9	5	1	4	8	3
5	8	4	3	2	7	1	9	6

INTERMEDIATE - 252

3	4	2	8	6	5	9	7	1
6	1	9	7	3	2	5	4	8
5	8	7	9	4	1	6	2	3
9	2	1	3	5	4	8	6	7
7	6	4	2	8	9	1	3	5
8	5	3	1	7	6	2	9	4
2	9	8	4	1	3	7	5	6
1	3	6	5	2	7	4	8	9
4	7	5	6	9	8	3	1	2

INTERMEDIATE - 253

8	9	7	1	5	2	3	6	4
6	1	4	3	9	7	5	2	8
2	5	3	8	6	4	1	9	7
5	4	8	2	3	6	9	7	1
9	3	6	5	7	1	4	8	2
1	7	2	4	8	9	6	5	3
7	8	5	9	4	3	2	1	6
4	2	9	6	1	8	7	3	5
3	6	1	7	2	5	8	4	9

INTERMEDIATE - 254

2	8	9	4	1	3	5	7	6
3	1	5	6	2	7	9	8	4
4	7	6	5	8	9	1	3	2
8	3	7	2	6	5	4	9	1
1	9	2	3	4	8	6	5	7
5	6	4	9	7	1	3	2	8
7	5	8	1	9	6	2	4	3
9	2	1	7	3	4	8	6	5
6	4	3	8	5	2	7	1	9

INTERMEDIATE - 255

9	5	2	7	4	6	3	8	1
8	7	6	3	2	1	4	5	9
3	4	1	8	9	5	6	7	2
2	1	4	6	7	9	5	3	8
5	6	8	2	3	4	1	9	7
7	3	9	5	1	8	2	4	6
1	2	5	4	8	7	9	6	3
6	8	3	9	5	2	7	1	4
4	9	7	1	6	3	8	2	5

INTERMEDIATE - 256

2	7	9	3	1	4	8	6	5
5	4	8	6	9	2	1	7	3
6	1	3	8	5	7	9	4	2
9	3	6	5	2	1	4	8	7
8	5	7	4	3	6	2	1	9
4	2	1	7	8	9	5	3	6
1	8	2	9	7	3	6	5	4
3	9	4	1	6	5	7	2	8
7	6	5	2	4	8	3	9	1

INTERMEDIATE - 257

5	2	3	1	8	6	4	9	7
4	6	7	5	9	2	1	3	8
1	8	9	4	7	3	2	6	5
7	4	5	6	3	1	9	8	2
3	1	6	8	2	9	5	7	4
8	9	2	7	4	5	6	1	3
9	7	1	2	5	8	3	4	6
6	5	8	3	1	4	7	2	9
2	3	4	9	6	7	8	5	1

INTERMEDIATE - 258

6	3	9	4	5	1	2	7	8
1	2	7	8	6	3	9	4	5
4	8	5	7	2	9	6	3	1
2	1	3	5	8	7	4	9	6
8	7	6	1	9	4	5	2	3
5	9	4	6	3	2	8	1	7
7	5	2	9	1	8	3	6	4
9	4	8	3	7	6	1	5	2
3	6	1	2	4	5	7	8	9

INTERMEDIATE - 259

5	3	2	9	6	7	8	4	1
8	7	1	4	3	2	5	6	9
6	4	9	1	8	5	7	3	2
1	8	4	5	2	3	9	7	6
3	6	5	7	9	1	2	8	4
9	2	7	6	4	8	1	5	3
7	9	6	2	5	4	3	1	8
4	1	8	3	7	9	6	2	5
2	5	3	8	1	6	4	9	7

INTERMEDIATE - 260

9	4	6	2	3	5	8	1	7
3	8	1	7	6	4	9	2	5
7	2	5	1	9	8	3	4	6
1	6	2	9	4	3	5	7	8
8	9	3	5	7	2	4	6	1
5	7	4	6	8	1	2	3	9
4	3	7	8	5	6	1	9	2
6	1	8	4	2	9	7	5	3
2	5	9	3	1	7	6	8	4

INTERMEDIATE - 261

6	1	9	4	5	7	2	3	8
4	7	3	2	8	1	9	6	5
2	8	5	6	9	3	1	7	4
8	4	7	9	1	6	3	5	2
1	3	6	5	2	8	4	9	7
5	9	2	7	3	4	6	8	1
3	6	8	1	4	5	7	2	9
9	5	1	3	7	2	8	4	6
7	2	4	8	6	9	5	1	3

INTERMEDIATE - 262

7	3	5	1	9	6	4	2	8
1	4	9	2	8	7	5	3	6
2	8	6	3	4	5	9	1	7
8	5	1	7	3	9	6	4	2
4	6	3	8	5	2	7	9	1
9	7	2	6	1	4	8	5	3
5	2	7	4	6	3	1	8	9
6	1	4	9	2	8	3	7	5
3	9	8	5	7	1	2	6	4

INTERMEDIATE - 263

5	6	1	4	2	3	7	8	9
4	8	9	7	1	5	6	2	3
2	7	3	9	8	6	4	5	1
8	9	7	2	3	4	5	1	6
3	1	4	5	6	7	8	9	2
6	5	2	1	9	8	3	4	7
9	3	6	8	5	1	2	7	4
7	2	8	3	4	9	1	6	5
1	4	5	6	7	2	9	3	8

INTERMEDIATE - 264

2	3	1	7	9	8	4	5	6
8	9	5	1	6	4	7	3	2
6	7	4	3	5	2	1	8	9
5	1	2	6	8	7	3	9	4
7	4	8	9	3	1	6	2	5
3	6	9	4	2	5	8	1	7
1	2	3	5	7	6	9	4	8
4	8	7	2	1	9	5	6	3
9	5	6	8	4	3	2	7	1

INTERMEDIATE - 265

5	9	7	8	3	2	1	6	4
2	4	3	6	1	5	7	9	8
1	8	6	7	4	9	2	5	3
8	6	9	4	2	1	3	7	5
4	2	5	9	7	3	8	1	6
7	3	1	5	6	8	4	2	9
9	5	2	1	8	4	6	3	7
3	7	4	2	5	6	9	8	1
6	1	8	3	9	7	5	4	2

INTERMEDIATE - 266

7	1	4	6	3	2	9	5	8
5	6	3	8	4	9	7	2	1
9	2	8	1	5	7	6	3	4
3	8	1	2	7	6	4	9	5
2	5	9	4	8	1	3	7	6
6	4	7	5	9	3	1	8	2
1	7	2	3	6	8	5	4	9
4	3	6	9	2	5	8	1	7
8	9	5	7	1	4	2	6	3

INTERMEDIATE - 267

9	2	7	1	8	3	6	5	4
1	4	6	2	5	9	8	7	3
8	3	5	4	7	6	2	1	9
7	1	4	6	3	5	9	2	8
2	5	9	7	4	8	3	6	1
6	8	3	9	2	1	5	4	7
3	9	1	5	6	4	7	8	2
5	7	8	3	1	2	4	9	6
4	6	2	8	9	7	1	3	5

INTERMEDIATE - 268

6	5	9	8	4	2	3	7	1
2	3	4	1	7	9	5	6	8
7	8	1	6	5	3	4	9	2
9	1	5	7	3	6	8	2	4
4	7	3	5	2	8	9	1	6
8	6	2	4	9	1	7	5	3
1	4	8	9	6	5	2	3	7
5	2	7	3	1	4	6	8	9
3	9	6	2	8	7	1	4	5

INTERMEDIATE -269

4	3	9	7	8	1	6	2	5
8	5	2	3	6	4	9	1	7
1	7	6	9	2	5	4	3	8
5	9	1	8	4	2	7	6	3
3	8	4	5	7	6	2	9	1
6	2	7	1	3	9	8	5	4
9	4	3	2	5	7	1	8	6
7	1	5	6	9	8	3	4	2
2	6	8	4	1	3	5	7	9

INTERMEDIATE - 270

1	7	4	9	6	3	8	2	5
5	3	8	2	4	7	9	6	1
9	2	6	8	5	1	7	4	3
2	1	3	5	7	6	4	9	8
4	9	7	1	8	2	5	3	6
8	6	5	3	9	4	1	7	2
3	8	9	4	2	5	6	1	7
6	5	1	7	3	9	2	8	4
7	4	2	6	1	8	3	5	9

INTERMEDIATE - 271

9	3	6	2	1	8	5	7	4
2	8	5	7	3	4	6	1	9
1	4	7	6	9	5	3	2	8
7	9	4	8	5	6	2	3	1
5	1	3	9	2	7	8	4	6
8	6	2	1	4	3	7	9	5
6	2	9	3	8	1	4	5	7
3	5	8	4	7	9	1	6	2
4	7	1	5	6	2	9	8	3

INTERMEDIATE - 272

3	4	9	8	7	2	6	5	1
2	8	6	3	5	1	7	4	9
1	7	5	9	6	4	2	8	3
5	1	2	6	4	9	3	7	8
7	9	3	2	8	5	1	6	4
8	6	4	1	3	7	9	2	5
6	3	7	4	9	8	5	1	2
4	5	1	7	2	3	8	9	6
9	2	8	5	1	6	4	3	7

INTERMEDIATE - 273

1	8	5	2	7	6	3	9	4
4	6	7	8	9	3	5	2	1
3	2	9	4	1	5	6	8	7
5	9	4	1	6	8	7	3	2
7	1	2	3	4	9	8	5	6
6	3	8	5	2	7	1	4	9
2	4	3	6	5	1	9	7	8
8	7	1	9	3	4	2	6	5
9	5	6	7	8	2	4	1	3

INTERMEDIATE - 274

8	4	1	7	2	5	9	3	6
7	3	6	8	9	1	2	5	4
9	5	2	3	6	4	7	1	8
1	9	7	6	3	8	5	4	2
2	6	4	1	5	7	3	8	9
5	8	3	9	4	2	6	7	1
4	1	9	2	7	3	8	6	5
6	7	5	4	8	9	1	2	3
3	2	8	5	1	6	4	9	7

INTERMEDIATE - 275

2	4	5	8	6	1	7	3	9
3	6	8	7	9	5	1	4	2
9	7	1	4	3	2	8	5	6
4	1	9	2	8	7	3	6	5
7	5	6	3	1	9	4	2	8
8	2	3	5	4	6	9	7	1
6	8	7	9	5	4	2	1	3
5	3	2	1	7	8	6	9	4
1	9	4	6	2	3	5	8	7

INTERMEDIATE - 276

1	7	5	9	2	3	6	8	4
6	9	2	8	4	7	1	5	3
4	8	3	1	5	6	7	2	9
3	1	4	7	9	2	5	6	8
7	2	8	6	3	5	9	4	1
9	5	6	4	8	1	3	7	2
2	6	9	3	7	4	8	1	5
8	4	1	5	6	9	2	3	7
5	3	7	2	1	8	4	9	6

INTERMEDIATE - 277

9	5	3	2	6	1	8	4	7
2	7	1	8	4	9	5	3	6
6	8	4	5	7	3	9	1	2
8	9	7	1	3	6	2	5	4
4	1	2	7	5	8	6	9	3
3	6	5	4	9	2	7	8	1
1	2	6	3	8	5	4	7	9
5	4	9	6	1	7	3	2	8
7	3	8	9	2	4	1	6	5

INTERMEDIATE - 278

5	1	4	7	3	2	6	9	8
3	2	9	6	4	8	1	5	7
8	6	7	9	1	5	2	3	4
9	7	3	8	5	6	4	2	1
1	8	6	4	2	3	9	7	5
2	4	5	1	9	7	8	6	3
4	5	1	2	7	9	3	8	6
6	3	2	5	8	1	7	4	9
7	9	8	3	6	4	5	1	2

INTERMEDIATE - 279

8	7	9	2	3	1	6	5	4
5	4	1	6	8	9	2	3	7
2	3	6	5	4	7	1	8	9
4	2	5	1	9	8	3	7	6
6	1	3	7	5	2	9	4	8
9	8	7	3	6	4	5	2	1
1	5	8	9	7	3	4	6	2
3	9	4	8	2	6	7	1	5
7	6	2	4	1	5	8	9	3

INTERMEDIATE - 280

6	2	9	5	7	3	8	4	1
3	8	1	2	6	4	7	9	5
7	5	4	1	8	9	2	3	6
5	9	8	7	3	2	6	1	4
2	3	6	4	5	1	9	7	8
4	1	7	6	9	8	5	2	3
8	7	2	3	1	6	4	5	9
1	6	5	9	4	7	3	8	2
9	4	3	8	2	5	1	6	7

INTERMEDIATE - 281

6	3	1	2	8	5	4	7	9
2	9	8	7	6	4	1	5	3
4	5	7	3	9	1	2	6	8
7	8	4	5	3	9	6	2	1
3	2	6	1	4	7	9	8	5
5	1	9	6	2	8	7	3	4
8	7	5	9	1	6	3	4	2
9	6	3	4	5	2	8	1	7
1	4	2	8	7	3	5	9	6

INTERMEDIATE - 282

5	4	2	1	3	6	8	9	7
6	7	8	9	2	4	3	5	1
9	3	1	7	5	8	6	4	2
8	9	5	3	7	1	2	6	4
3	1	6	4	9	2	5	7	8
7	2	4	6	8	5	1	3	9
1	8	7	5	6	9	4	2	3
2	5	3	8	4	7	9	1	6
4	6	9	2	1	3	7	8	5

INTERMEDIATE - 283

5	2	4	1	7	9	8	3	6
1	3	8	6	5	4	9	2	7
6	7	9	8	3	2	1	5	4
2	6	7	5	4	8	3	1	9
3	8	1	2	9	6	7	4	5
4	9	5	3	1	7	6	8	2
8	4	6	7	2	1	5	9	3
7	5	2	9	8	3	4	6	1
9	1	3	4	6	5	2	7	8

INTERMEDIATE - 284

1	8	6	5	4	7	3	2	9
9	7	5	3	8	2	1	4	6
4	2	3	6	1	9	8	7	5
6	9	1	7	5	3	4	8	2
3	4	2	8	9	6	5	1	7
8	5	7	4	2	1	6	9	3
5	6	9	1	7	8	2	3	4
2	1	4	9	3	5	7	6	8
7	3	8	2	6	4	9	5	1

INTERMEDIATE - 285

4	6	9	3	8	7	1	5	2
7	2	1	6	5	4	9	8	3
5	8	3	9	1	2	6	7	4
9	4	5	2	3	1	8	6	7
2	3	6	8	7	5	4	1	9
8	1	7	4	9	6	2	3	5
3	5	2	1	4	8	7	9	6
1	9	4	7	6	3	5	2	8
6	7	8	5	2	9	3	4	1

INTERMEDIATE - 286

1	3	6	5	7	9	8	4	2
8	4	7	1	2	3	9	6	5
9	2	5	8	6	4	3	1	7
4	7	8	2	1	5	6	3	9
2	6	3	9	8	7	4	5	1
5	1	9	3	4	6	7	2	8
3	5	1	4	9	8	2	7	6
6	8	4	7	5	2	1	9	3
7	9	2	6	3	1	5	8	4

INTERMEDIATE - 287

6	3	5	2	1	7	4	8	9
4	9	2	3	8	5	7	1	6
8	1	7	6	4	9	3	2	5
9	6	1	5	2	3	8	4	7
2	7	3	4	9	8	6	5	1
5	8	4	7	6	1	9	3	2
1	4	6	9	3	2	5	7	8
7	2	9	8	5	4	1	6	3
3	5	8	1	7	6	2	9	4

INTERMEDIATE - 288

9	4	5	6	1	2	7	3	8
3	7	2	4	5	8	1	6	9
8	1	6	3	9	7	2	4	5
1	8	3	5	2	6	9	7	4
2	5	9	7	3	4	6	8	1
7	6	4	1	8	9	3	5	2
5	3	7	2	4	1	8	9	6
4	9	1	8	6	3	5	2	7
6	2	8	9	7	5	4	1	3

INTERMEDIATE - 289

4	1	2	3	8	5	7	9	6
9	3	8	2	6	7	5	1	4
6	7	5	1	4	9	2	8	3
2	4	7	9	5	1	3	6	8
1	8	6	4	7	3	9	2	5
5	9	3	6	2	8	1	4	7
7	2	1	8	3	6	4	5	9
3	6	4	5	9	2	8	7	1
8	5	9	7	1	4	6	3	2

INTERMEDIATE - 290

7	5	6	4	2	8	9	1	3
1	3	2	5	9	7	6	8	4
9	8	4	3	6	1	7	2	5
3	1	8	6	5	9	2	4	7
6	2	7	1	8	4	3	5	9
4	9	5	2	7	3	8	6	1
2	7	9	8	1	5	4	3	6
8	4	1	9	3	6	5	7	2
5	6	3	7	4	2	1	9	8

INTERMEDIATE - 291

4	6	7	2	9	1	5	3	8
9	1	2	8	5	3	4	6	7
5	8	3	7	6	4	9	1	2
6	9	1	4	2	5	7	8	3
7	4	8	3	1	6	2	9	5
3	2	5	9	7	8	1	4	6
2	5	6	1	8	9	3	7	4
1	7	4	6	3	2	8	5	9
8	3	9	5	4	7	6	2	1

INTERMEDIATE - 292

8	2	6	9	7	5	3	1	4
3	9	5	1	8	4	7	6	2
4	7	1	6	2	3	8	9	5
5	6	3	8	4	7	9	2	1
9	4	7	2	6	1	5	8	3
1	8	2	3	5	9	6	4	7
7	1	9	4	3	8	2	5	6
6	5	8	7	1	2	4	3	9
2	3	4	5	9	6	1	7	8

INTERMEDIATE - 293

5	4	1	2	9	6	8	7	3
6	2	9	7	3	8	1	4	5
8	7	3	5	4	1	9	2	6
2	3	7	1	8	5	4	6	9
9	8	4	6	2	3	5	1	7
1	5	6	4	7	9	3	8	2
3	9	2	8	1	7	6	5	4
7	1	5	3	6	4	2	9	8
4	6	8	9	5	2	7	3	1

INTERMEDIATE - 294

8	5	7	2	6	4	3	9	1
4	3	2	7	1	9	8	5	6
1	6	9	5	3	8	2	4	7
9	1	6	8	7	5	4	3	2
2	4	3	6	9	1	7	8	5
7	8	5	4	2	3	1	6	9
5	2	1	3	4	6	9	7	8
3	7	8	9	5	2	6	1	4
6	9	4	1	8	7	5	2	3

INTERMEDIATE - 295

1	6	8	9	7	2	4	5	3
2	9	7	4	5	3	6	1	8
5	4	3	1	8	6	2	7	9
6	3	1	2	4	9	5	8	7
4	7	9	5	1	8	3	6	2
8	2	5	6	3	7	1	9	4
9	1	4	8	2	5	7	3	6
3	8	2	7	6	1	9	4	5
7	5	6	3	9	4	8	2	1

INTERMEDIATE - 296

2	1	6	5	3	7	4	8	9
5	4	3	9	2	8	7	6	1
9	7	8	6	4	1	5	3	2
8	5	7	1	9	4	6	2	3
6	2	1	3	7	5	9	4	8
3	9	4	8	6	2	1	7	5
4	6	5	2	8	9	3	1	7
1	3	2	7	5	6	8	9	4
7	8	9	4	1	3	2	5	6

INTERMEDIATE - 297

5	4	9	1	7	6	2	3	8
8	1	6	2	3	5	9	7	4
7	2	3	4	9	8	5	6	1
9	3	8	5	4	2	6	1	7
2	6	5	7	8	1	3	4	9
4	7	1	9	6	3	8	2	5
6	5	2	8	1	4	7	9	3
3	9	4	6	5	7	1	8	2
1	8	7	3	2	9	4	5	6

INTERMEDIATE - 298

6	8	2	1	3	9	7	5	4
9	3	1	5	4	7	6	8	2
4	7	5	8	6	2	9	3	1
5	6	7	9	1	3	2	4	8
1	4	3	2	8	6	5	9	7
2	9	8	7	5	4	1	6	3
8	5	9	3	2	1	4	7	6
3	1	6	4	7	5	8	2	9
7	2	4	6	9	8	3	1	5

INTERMEDIATE - 299

5	2	9	3	8	7	6	1	4
6	8	4	2	5	1	9	3	7
1	7	3	4	6	9	2	5	8
9	1	5	8	7	4	3	2	6
4	3	7	6	9	2	5	8	1
8	6	2	5	1	3	4	7	9
7	4	8	9	2	5	1	6	3
3	5	1	7	4	6	8	9	2
2	9	6	1	3	8	7	4	5

INTERMEDIATE - 300

3	7	6	4	9	8	2	1	5
1	8	2	7	5	3	9	6	4
4	5	9	2	1	6	3	7	8
9	2	5	6	4	7	1	8	3
6	1	4	3	8	5	7	2	9
8	3	7	1	2	9	4	5	6
2	9	3	8	6	1	5	4	7
7	6	1	5	3	4	8	9	2
5	4	8	9	7	2	6	3	1

SUDOKU
SOLUTIONS
EXPERT - {1 -500}

EXPERT - 01

7	4	3	8	1	9	6	5	2
2	6	9	3	7	5	4	8	1
8	1	5	4	6	2	3	7	9
6	7	4	1	9	8	5	2	3
3	8	2	6	5	4	1	9	7
5	9	1	7	2	3	8	4	6
1	5	7	2	4	6	9	3	8
9	3	6	5	8	7	2	1	4
4	2	8	9	3	1	7	6	5

EXPERT - 02

3	1	8	2	5	7	9	4	6
2	9	7	4	6	1	3	5	8
6	5	4	8	9	3	1	7	2
4	7	5	3	8	2	6	9	1
1	3	9	7	4	6	2	8	5
8	6	2	5	1	9	4	3	7
7	4	1	9	2	5	8	6	3
9	2	3	6	7	8	5	1	4
5	8	6	1	3	4	7	2	9

EXPERT - 03

5	8	6	9	2	3	1	7	4
7	4	9	5	1	6	3	8	2
3	2	1	8	4	7	5	6	9
9	7	2	4	5	8	6	3	1
1	3	8	7	6	9	2	4	5
6	5	4	2	3	1	8	9	7
2	6	5	3	7	4	9	1	8
8	1	7	6	9	5	4	2	3
4	9	3	1	8	2	7	5	6

EXPERT - 04

5	1	9	3	8	7	4	6	2
7	3	4	2	6	1	9	5	8
8	2	6	5	9	4	3	1	7
1	6	7	9	3	8	5	2	4
4	5	3	6	7	2	1	8	9
9	8	2	4	1	5	7	3	6
3	4	5	8	2	9	6	7	1
2	9	1	7	5	6	8	4	3
6	7	8	1	4	3	2	9	5

EXPERT - 05

5	6	1	4	8	7	3	2	9
3	4	8	2	9	6	7	5	1
2	9	7	5	3	1	4	8	6
9	1	6	3	4	8	5	7	2
8	5	4	7	1	2	9	6	3
7	3	2	6	5	9	8	1	4
1	8	3	9	2	5	6	4	7
6	2	9	8	7	4	1	3	5
4	7	5	1	6	3	2	9	8

EXPERT - 06

9	4	8	6	2	1	3	7	5
2	3	6	7	4	5	8	1	9
5	1	7	9	3	8	2	6	4
7	9	5	4	8	2	6	3	1
4	6	3	1	7	9	5	2	8
1	8	2	3	5	6	9	4	7
6	5	9	2	1	4	7	8	3
3	2	4	8	9	7	1	5	6
8	7	1	5	6	3	4	9	2

EXPERT - 07

2	9	5	6	3	4	7	1	8
1	8	4	7	2	9	6	3	5
3	7	6	5	8	1	4	9	2
8	6	3	1	4	7	2	5	9
7	5	2	9	6	3	8	4	1
9	4	1	8	5	2	3	7	6
6	2	7	4	9	5	1	8	3
5	1	8	3	7	6	9	2	4
4	3	9	2	1	8	5	6	7

EXPERT - 08

5	3	6	8	7	1	4	2	9
7	2	4	6	9	5	1	3	8
9	1	8	2	3	4	5	7	6
1	5	7	4	8	3	9	6	2
2	8	3	1	6	9	7	4	5
4	6	9	7	5	2	3	8	1
8	9	5	3	4	6	2	1	7
3	7	1	9	2	8	6	5	4
6	4	2	5	1	7	8	9	3

EXPERT - 09

6	9	5	2	1	3	7	8	4
8	4	1	6	9	7	3	2	5
7	2	3	8	5	4	6	9	1
1	8	6	7	2	9	4	5	3
4	3	2	5	6	8	1	7	9
5	7	9	3	4	1	8	6	2
9	5	8	1	3	6	2	4	7
3	6	4	9	7	2	5	1	8
2	1	7	4	8	5	9	3	6

EXPERT - 10

7	1	9	3	6	4	8	2	5
2	3	8	9	1	5	7	6	4
4	6	5	7	8	2	3	1	9
5	4	2	1	7	8	6	9	3
8	7	6	5	9	3	1	4	2
1	9	3	2	4	6	5	8	7
3	5	1	8	2	9	4	7	6
6	2	7	4	3	1	9	5	8
9	8	4	6	5	7	2	3	1

EXPERT - 11

5	6	7	3	8	1	2	9	4
1	4	3	2	7	9	6	5	8
8	9	2	5	6	4	1	7	3
6	3	9	8	4	7	5	2	1
4	5	1	9	2	6	8	3	7
7	2	8	1	5	3	9	4	6
3	7	5	6	9	8	4	1	2
9	1	6	4	3	2	7	8	5
2	8	4	7	1	5	3	6	9

EXPERT - 12

2	6	4	7	3	5	1	8	9
7	8	9	6	1	4	5	2	3
5	1	3	9	8	2	4	7	6
8	7	2	4	5	9	3	6	1
3	9	5	8	6	1	2	4	7
6	4	1	3	2	7	8	9	5
9	5	6	1	4	8	7	3	2
1	3	8	2	7	6	9	5	4
4	2	7	5	9	3	6	1	8

EXPERT - 13

9	2	1	5	4	6	7	8	3
3	5	8	1	7	2	9	6	4
6	7	4	8	3	9	5	2	1
7	9	2	4	6	1	3	5	8
4	3	6	2	5	8	1	9	7
8	1	5	3	9	7	6	4	2
5	4	7	6	2	3	8	1	9
1	6	3	9	8	4	2	7	5
2	8	9	7	1	5	4	3	6

EXPERT - 14

7	9	6	2	1	4	3	5	8
5	1	4	6	8	3	7	9	2
3	8	2	9	5	7	4	6	1
6	3	1	8	7	2	5	4	9
8	2	9	4	3	5	1	7	6
4	7	5	1	6	9	8	2	3
1	4	3	5	2	6	9	8	7
2	5	8	7	9	1	6	3	4
9	6	7	3	4	8	2	1	5

EXPERT - 15

3	5	4	2	1	8	6	9	7
9	1	8	5	6	7	2	4	3
6	7	2	9	3	4	8	5	1
4	2	5	8	9	3	7	1	6
7	6	1	4	2	5	3	8	9
8	3	9	1	7	6	5	2	4
5	8	6	3	4	9	1	7	2
1	4	3	7	5	2	9	6	8
2	9	7	6	8	1	4	3	5

EXPERT - 16

4	7	8	5	6	9	3	2	1
1	2	6	8	3	7	9	4	5
5	9	3	4	2	1	6	7	8
6	1	7	9	8	2	5	3	4
8	3	4	1	5	6	7	9	2
2	5	9	7	4	3	8	1	6
7	4	5	3	1	8	2	6	9
3	8	2	6	9	4	1	5	7
9	6	1	2	7	5	4	8	3

EXPERT - 17

1	8	7	3	5	9	6	2	4
2	6	9	7	4	8	5	3	1
5	3	4	2	6	1	9	7	8
4	7	2	6	9	3	1	8	5
8	9	6	5	1	7	2	4	3
3	1	5	4	8	2	7	9	6
7	5	3	1	2	4	8	6	9
9	4	1	8	7	6	3	5	2
6	2	8	9	3	5	4	1	7

EXPERT - 18

4	5	1	7	2	9	6	8	3
2	3	9	8	1	6	5	4	7
6	8	7	4	5	3	2	9	1
7	4	5	6	8	1	9	3	2
9	2	3	5	4	7	1	6	8
1	6	8	3	9	2	4	7	5
8	1	2	9	3	4	7	5	6
5	7	4	1	6	8	3	2	9
3	9	6	2	7	5	8	1	4

EXPERT - 19

1	5	2	4	3	8	7	6	9
3	4	8	9	7	6	5	1	2
9	7	6	2	5	1	3	4	8
7	2	9	1	6	3	8	5	4
8	1	4	5	2	7	6	9	3
5	6	3	8	9	4	2	7	1
2	8	5	7	4	9	1	3	6
6	9	1	3	8	5	4	2	7
4	3	7	6	1	2	9	8	5

EXPERT - 20

4	1	7	9	3	5	2	6	8
8	3	6	1	7	2	9	5	4
5	2	9	8	6	4	1	7	3
6	4	8	7	2	1	3	9	5
9	5	1	3	4	6	7	8	2
2	7	3	5	8	9	6	4	1
1	6	4	2	9	8	5	3	7
7	8	5	6	1	3	4	2	9
3	9	2	4	5	7	8	1	6

EXPERT - 21

4	5	2	8	3	9	1	6	7
8	7	3	5	6	1	4	9	2
1	9	6	4	7	2	3	5	8
7	4	8	9	1	3	6	2	5
9	6	5	2	4	8	7	1	3
2	3	1	7	5	6	9	8	4
5	1	4	6	2	7	8	3	9
6	2	9	3	8	4	5	7	1
3	8	7	1	9	5	2	4	6

EXPERT - 22

1	3	9	4	5	6	2	7	8
7	8	6	9	2	1	4	5	3
5	4	2	3	7	8	1	6	9
9	7	3	1	6	5	8	4	2
8	5	4	2	9	7	3	1	6
6	2	1	8	3	4	5	9	7
4	9	8	6	1	3	7	2	5
3	6	7	5	4	2	9	8	1
2	1	5	7	8	9	6	3	4

EXPERT - 23

5	1	7	6	3	2	9	8	4
9	4	3	8	7	1	6	5	2
6	2	8	5	9	4	1	7	3
3	8	1	2	5	7	4	9	6
2	5	4	9	8	6	3	1	7
7	6	9	1	4	3	8	2	5
1	7	2	3	6	8	5	4	9
4	9	6	7	1	5	2	3	8
8	3	5	4	2	9	7	6	1

EXPERT - 24

5	3	4	9	8	1	6	7	2
2	1	6	5	4	7	9	8	3
9	7	8	2	3	6	4	5	1
4	2	7	6	9	3	8	1	5
8	5	9	7	1	2	3	4	6
1	6	3	8	5	4	2	9	7
6	8	1	3	7	9	5	2	4
3	4	5	1	2	8	7	6	9
7	9	2	4	6	5	1	3	8

EXPERT - 25

5	9	7	8	6	3	1	2	4
6	1	2	5	4	9	8	7	3
4	8	3	7	1	2	5	9	6
2	3	9	1	7	6	4	5	8
1	5	6	4	2	8	7	3	9
7	4	8	3	9	5	2	6	1
3	2	1	6	8	7	9	4	5
9	6	4	2	5	1	3	8	7
8	7	5	9	3	4	6	1	2

EXPERT - 26

7	1	5	3	6	4	2	9	8
6	2	9	5	8	7	3	4	1
3	4	8	2	1	9	5	6	7
9	8	7	1	2	5	4	3	6
1	5	3	7	4	6	8	2	9
4	6	2	8	9	3	7	1	5
2	9	4	6	5	8	1	7	3
5	3	1	9	7	2	6	8	4
8	7	6	4	3	1	9	5	2

EXPERT - 27

3	8	5	2	1	4	6	7	9
7	1	9	5	8	6	3	4	2
4	2	6	7	3	9	8	1	5
5	3	2	8	9	7	1	6	4
6	9	8	4	2	1	7	5	3
1	7	4	6	5	3	9	2	8
2	4	1	3	7	8	5	9	6
8	6	7	9	4	5	2	3	1
9	5	3	1	6	2	4	8	7

EXPERT - 28

8	3	2	5	6	4	1	9	7
4	9	5	8	7	1	6	3	2
7	6	1	9	3	2	8	4	5
5	8	3	6	2	7	9	1	4
6	1	7	3	4	9	5	2	8
9	2	4	1	5	8	7	6	3
1	5	6	2	8	3	4	7	9
2	7	8	4	9	6	3	5	1
3	4	9	7	1	5	2	8	6

EXPERT - 29

6	9	3	7	5	2	1	4	8
7	1	4	3	6	8	9	5	2
2	5	8	1	9	4	7	3	6
3	2	9	8	7	5	6	1	4
8	4	1	6	2	3	5	9	7
5	6	7	4	1	9	8	2	3
4	3	5	9	8	7	2	6	1
1	7	2	5	4	6	3	8	9
9	8	6	2	3	1	4	7	5

EXPERT - 30

2	3	4	8	1	5	6	9	7
9	1	8	6	2	7	3	5	4
5	6	7	3	9	4	8	2	1
1	4	6	9	7	3	5	8	2
3	9	2	5	8	1	7	4	6
8	7	5	2	4	6	9	1	3
7	5	9	4	6	2	1	3	8
4	8	1	7	3	9	2	6	5
6	2	3	1	5	8	4	7	9

EXPERT - 31

9	6	1	7	8	5	3	4	2
4	2	5	6	1	3	8	9	7
7	3	8	2	4	9	6	5	1
6	9	3	8	5	7	2	1	4
5	7	4	1	3	2	9	6	8
8	1	2	4	9	6	7	3	5
1	5	7	3	6	8	4	2	9
2	4	6	9	7	1	5	8	3
3	8	9	5	2	4	1	7	6

EXPERT - 32

1	3	4	5	9	6	8	7	2
8	5	6	2	7	3	9	4	1
2	9	7	1	8	4	6	3	5
7	4	2	3	6	8	1	5	9
5	8	3	9	1	2	7	6	4
9	6	1	4	5	7	3	2	8
3	1	8	7	4	5	2	9	6
6	7	5	8	2	9	4	1	3
4	2	9	6	3	1	5	8	7

EXPERT - 33

9	7	8	2	4	6	3	1	5
4	3	1	8	5	7	2	6	9
5	6	2	1	9	3	4	8	7
7	1	5	4	8	2	6	9	3
3	2	6	5	7	9	1	4	8
8	4	9	6	3	1	5	7	2
2	9	7	3	6	4	8	5	1
6	8	3	7	1	5	9	2	4
1	5	4	9	2	8	7	3	6

EXPERT - 34

6	5	1	3	7	8	2	9	4
7	2	9	5	1	4	6	8	3
3	8	4	6	9	2	1	7	5
8	7	5	1	4	9	3	6	2
9	1	6	8	2	3	5	4	7
4	3	2	7	6	5	9	1	8
5	9	7	4	3	1	8	2	6
2	6	8	9	5	7	4	3	1
1	4	3	2	8	6	7	5	9

EXPERT - 35

8	4	1	6	2	7	3	9	5
9	3	6	8	5	4	2	7	1
2	5	7	3	9	1	8	6	4
6	7	9	5	3	8	1	4	2
5	1	2	4	6	9	7	8	3
4	8	3	1	7	2	6	5	9
3	6	8	9	1	5	4	2	7
7	9	4	2	8	3	5	1	6
1	2	5	7	4	6	9	3	8

EXPERT - 36

5	2	9	8	6	4	3	1	7
3	7	8	2	1	5	6	9	4
4	1	6	3	7	9	5	2	8
6	8	7	5	9	1	2	4	3
9	5	2	6	4	3	8	7	1
1	3	4	7	8	2	9	5	6
8	4	1	9	2	6	7	3	5
7	9	5	4	3	8	1	6	2
2	6	3	1	5	7	4	8	9

EXPERT - 37

7	5	1	9	8	3	6	2	4
3	8	6	4	7	2	5	1	9
9	2	4	1	5	6	8	7	3
5	4	3	2	6	8	7	9	1
2	6	9	7	1	5	3	4	8
8	1	7	3	9	4	2	6	5
1	3	2	5	4	7	9	8	6
4	7	8	6	3	9	1	5	2
6	9	5	8	2	1	4	3	7

EXPERT - 38

1	6	7	5	9	8	4	2	3
5	4	8	2	7	3	1	9	6
2	3	9	1	6	4	5	7	8
4	7	1	3	8	2	6	5	9
6	8	3	9	1	5	2	4	7
9	5	2	7	4	6	3	8	1
7	9	6	4	2	1	8	3	5
8	2	5	6	3	9	7	1	4
3	1	4	8	5	7	9	6	2

EXPERT - 39

2	6	8	9	7	4	5	3	1
1	9	7	5	3	2	4	8	6
3	4	5	8	1	6	7	9	2
5	1	2	7	8	9	6	4	3
9	7	3	6	4	1	8	2	5
4	8	6	2	5	3	1	7	9
6	5	4	3	9	8	2	1	7
7	3	1	4	2	5	9	6	8
8	2	9	1	6	7	3	5	4

EXPERT - 40

4	9	3	5	7	2	6	8	1
5	7	8	6	1	4	3	9	2
2	6	1	8	3	9	4	5	7
3	5	9	7	4	6	1	2	8
7	8	4	9	2	1	5	3	6
6	1	2	3	8	5	7	4	9
1	4	6	2	9	3	8	7	5
9	3	7	1	5	8	2	6	4
8	2	5	4	6	7	9	1	3

EXPERT - 41

5	6	3	1	4	2	8	9	7
9	1	4	5	7	8	3	2	6
2	7	8	9	3	6	5	4	1
6	3	2	7	5	1	4	8	9
1	5	7	8	9	4	6	3	2
4	8	9	2	6	3	7	1	5
8	2	6	4	1	5	9	7	3
3	9	1	6	8	7	2	5	4
7	4	5	3	2	9	1	6	8

EXPERT - 42

6	4	3	5	7	9	8	2	1
8	9	2	3	1	4	5	7	6
5	1	7	6	2	8	4	9	3
1	2	9	4	8	5	3	6	7
3	6	8	2	9	7	1	5	4
7	5	4	1	3	6	2	8	9
4	3	5	7	6	2	9	1	8
9	7	1	8	5	3	6	4	2
2	8	6	9	4	1	7	3	5

EXPERT - 43

7	6	1	5	8	2	4	3	9
4	5	2	7	3	9	1	6	8
3	9	8	1	4	6	5	2	7
9	7	5	8	6	4	3	1	2
1	4	6	2	7	3	9	8	5
2	8	3	9	5	1	7	4	6
5	3	7	6	1	8	2	9	4
6	2	4	3	9	5	8	7	1
8	1	9	4	2	7	6	5	3

EXPERT - 44

8	1	2	3	5	6	4	7	9
5	6	9	7	4	2	8	1	3
4	7	3	8	1	9	2	6	5
3	2	8	1	6	4	9	5	7
9	5	6	2	7	3	1	4	8
7	4	1	9	8	5	6	3	2
1	8	4	5	9	7	3	2	6
6	3	7	4	2	8	5	9	1
2	9	5	6	3	1	7	8	4

EXPERT - 45

1	4	6	8	7	2	9	5	3
7	5	8	3	9	4	2	6	1
3	2	9	1	5	6	8	4	7
2	6	7	4	1	8	5	3	9
8	3	5	9	2	7	6	1	4
4	9	1	5	6	3	7	2	8
6	1	2	7	4	9	3	8	5
5	7	3	6	8	1	4	9	2
9	8	4	2	3	5	1	7	6

EXPERT - 46

9	1	3	8	4	2	5	7	6
6	2	4	5	7	1	3	8	9
8	5	7	9	3	6	1	2	4
5	4	9	1	2	3	8	6	7
1	7	2	4	6	8	9	3	5
3	6	8	7	5	9	4	1	2
4	3	1	6	9	7	2	5	8
7	8	5	2	1	4	6	9	3
2	9	6	3	8	5	7	4	1

EXPERT - 47

8	7	4	3	6	2	1	5	9
9	3	2	1	7	5	6	8	4
1	5	6	9	4	8	7	2	3
6	2	1	8	3	4	9	7	5
7	4	5	2	1	9	8	3	6
3	8	9	7	5	6	4	1	2
4	1	8	5	9	3	2	6	7
5	6	7	4	2	1	3	9	8
2	9	3	6	8	7	5	4	1

EXPERT - 48

8	2	6	3	5	1	7	9	4
5	9	3	7	8	4	2	6	1
4	7	1	6	9	2	5	8	3
6	1	9	8	7	3	4	2	5
2	4	8	1	6	5	3	7	9
3	5	7	2	4	9	8	1	6
1	3	5	9	2	7	6	4	8
7	6	4	5	1	8	9	3	2
9	8	2	4	3	6	1	5	7

EXPERT - 49

8	7	3	4	6	5	2	1	9
6	4	1	3	2	9	5	7	8
9	5	2	7	8	1	3	6	4
3	9	5	6	1	4	8	2	7
2	8	7	5	9	3	6	4	1
1	6	4	8	7	2	9	3	5
7	2	9	1	3	8	4	5	6
4	1	8	2	5	6	7	9	3
5	3	6	9	4	7	1	8	2

EXPERT - 50

3	2	6	7	5	8	1	4	9
7	4	8	3	1	9	5	6	2
9	1	5	4	6	2	7	3	8
8	7	3	2	9	6	4	5	1
6	5	2	1	4	3	9	8	7
1	9	4	8	7	5	3	2	6
5	6	7	9	8	4	2	1	3
2	8	1	5	3	7	6	9	4
4	3	9	6	2	1	8	7	5

EXPERT - 51

3	4	7	5	2	1	9	8	6
1	5	8	9	7	6	3	4	2
2	6	9	8	4	3	5	7	1
7	8	6	3	1	2	4	5	9
9	2	5	4	6	8	1	3	7
4	1	3	7	5	9	6	2	8
8	9	2	6	3	4	7	1	5
6	7	4	1	8	5	2	9	3
5	3	1	2	9	7	8	6	4

EXPERT - 52

1	5	9	6	2	3	7	8	4
3	8	2	4	7	5	6	9	1
7	6	4	8	1	9	2	5	3
6	1	8	7	3	2	5	4	9
4	9	3	1	5	6	8	2	7
2	7	5	9	4	8	1	3	6
8	3	6	2	9	1	4	7	5
9	4	1	5	8	7	3	6	2
5	2	7	3	6	4	9	1	8

EXPERT - 53

1	4	8	5	7	9	2	6	3
2	5	7	1	6	3	9	4	8
6	9	3	2	4	8	7	5	1
7	8	6	3	1	4	5	9	2
3	1	4	9	5	2	6	8	7
5	2	9	7	8	6	1	3	4
8	7	1	4	9	5	3	2	6
4	3	5	6	2	7	8	1	9
9	6	2	8	3	1	4	7	5

EXPERT - 54

4	6	8	9	5	1	7	3	2
3	1	2	7	6	8	5	9	4
5	7	9	3	2	4	1	8	6
8	5	3	6	4	7	9	2	1
1	9	6	5	3	2	4	7	8
7	2	4	1	8	9	3	6	5
9	4	5	2	7	6	8	1	3
2	8	1	4	9	3	6	5	7
6	3	7	8	1	5	2	4	9

EXPERT - 55

5	9	7	8	2	1	3	6	4
8	2	6	7	3	4	9	1	5
1	3	4	6	5	9	8	7	2
6	8	2	9	4	3	1	5	7
7	4	1	5	8	2	6	3	9
9	5	3	1	7	6	4	2	8
2	6	8	4	1	5	7	9	3
3	7	9	2	6	8	5	4	1
4	1	5	3	9	7	2	8	6

EXPERT - 56

2	1	8	4	6	5	3	7	9
4	5	6	7	3	9	1	2	8
9	7	3	8	2	1	4	6	5
8	3	4	5	9	6	2	1	7
1	2	7	3	4	8	9	5	6
6	9	5	1	7	2	8	3	4
5	4	1	6	8	3	7	9	2
3	8	9	2	5	7	6	4	1
7	6	2	9	1	4	5	8	3

EXPERT - 57

3	6	4	9	5	1	2	8	7
7	9	5	4	2	8	1	3	6
1	2	8	6	7	3	5	9	4
9	4	3	5	6	2	7	1	8
5	7	1	3	8	9	4	6	2
6	8	2	7	1	4	9	5	3
8	1	6	2	9	7	3	4	5
4	5	7	1	3	6	8	2	9
2	3	9	8	4	5	6	7	1

EXPERT - 58

8	2	7	5	4	3	1	6	9
1	3	5	9	7	6	2	4	8
6	9	4	2	1	8	5	7	3
5	6	2	1	8	7	3	9	4
4	8	9	3	2	5	6	1	7
3	7	1	6	9	4	8	2	5
9	4	3	8	6	1	7	5	2
7	1	8	4	5	2	9	3	6
2	5	6	7	3	9	4	8	1

EXPERT - 59

9	6	7	4	1	3	2	8	5
4	5	8	2	6	7	1	9	3
3	2	1	9	5	8	6	4	7
7	9	5	1	3	2	8	6	4
1	4	6	5	8	9	7	3	2
8	3	2	7	4	6	9	5	1
2	7	3	8	9	4	5	1	6
5	8	4	6	2	1	3	7	9
6	1	9	3	7	5	4	2	8

EXPERT - 60

8	2	3	1	7	5	4	6	9
6	5	1	9	8	4	2	7	3
9	4	7	3	6	2	1	8	5
3	6	2	7	9	1	8	5	4
4	8	9	5	2	6	7	3	1
1	7	5	4	3	8	6	9	2
5	3	8	2	4	7	9	1	6
2	9	6	8	1	3	5	4	7
7	1	4	6	5	9	3	2	8

EXPERT - 61

5	1	6	2	4	3	7	8	9
3	2	8	9	6	7	5	1	4
4	7	9	1	5	8	3	6	2
6	5	7	4	8	9	1	2	3
9	3	1	7	2	5	8	4	6
2	8	4	6	3	1	9	7	5
8	4	5	3	7	2	6	9	1
7	9	2	5	1	6	4	3	8
1	6	3	8	9	4	2	5	7

EXPERT - 62

6	9	1	2	8	5	3	7	4
2	7	4	9	1	3	6	5	8
3	5	8	4	7	6	9	1	2
5	3	9	6	2	4	7	8	1
8	1	6	7	5	9	4	2	3
4	2	7	1	3	8	5	9	6
7	8	5	3	6	2	1	4	9
1	4	3	8	9	7	2	6	5
9	6	2	5	4	1	8	3	7

EXPERT - 63

8	9	2	5	3	6	1	7	4
4	6	5	1	7	9	2	8	3
1	7	3	2	4	8	9	6	5
6	2	4	8	9	1	5	3	7
9	8	1	7	5	3	6	4	2
5	3	7	4	6	2	8	1	9
2	4	6	3	8	5	7	9	1
3	5	8	9	1	7	4	2	6
7	1	9	6	2	4	3	5	8

EXPERT - 64

5	4	2	8	9	3	6	1	7
1	3	9	2	7	6	4	8	5
7	6	8	1	4	5	3	2	9
6	8	3	7	5	2	9	4	1
9	5	1	6	8	4	2	7	3
2	7	4	9	3	1	8	5	6
4	1	6	5	2	9	7	3	8
8	2	5	3	6	7	1	9	4
3	9	7	4	1	8	5	6	2

EXPERT - 65

5	3	1	6	7	4	2	8	9
6	4	8	2	5	9	7	3	1
2	7	9	1	3	8	6	5	4
9	8	6	7	1	5	3	4	2
4	1	2	8	6	3	9	7	5
7	5	3	4	9	2	8	1	6
8	9	4	5	2	7	1	6	3
1	2	5	3	8	6	4	9	7
3	6	7	9	4	1	5	2	8

EXPERT - 66

3	1	9	5	7	2	8	6	4
6	2	8	1	4	3	5	7	9
5	4	7	9	8	6	1	2	3
9	8	6	2	3	1	4	5	7
4	3	5	6	9	7	2	8	1
1	7	2	4	5	8	9	3	6
7	5	4	8	6	9	3	1	2
8	6	1	3	2	4	7	9	5
2	9	3	7	1	5	6	4	8

EXPERT - 67

7	9	4	5	6	1	2	8	3
8	3	2	9	7	4	6	1	5
6	1	5	3	2	8	9	7	4
9	8	6	7	5	2	3	4	1
5	4	3	1	8	6	7	2	9
2	7	1	4	3	9	5	6	8
3	6	8	2	1	5	4	9	7
4	2	7	8	9	3	1	5	6
1	5	9	6	4	7	8	3	2

EXPERT - 68

5	4	2	1	9	3	7	8	6
9	7	6	8	2	4	1	5	3
3	1	8	5	7	6	9	4	2
7	2	3	4	6	1	8	9	5
6	5	1	9	8	7	3	2	4
8	9	4	3	5	2	6	1	7
4	8	7	2	3	9	5	6	1
2	3	5	6	1	8	4	7	9
1	6	9	7	4	5	2	3	8

EXPERT - 69

9	7	2	5	8	3	6	4	1
8	4	5	7	1	6	2	3	9
6	3	1	2	4	9	5	7	8
4	9	3	1	6	8	7	5	2
2	1	8	4	7	5	3	9	6
5	6	7	9	3	2	8	1	4
7	8	9	3	2	4	1	6	5
3	2	4	6	5	1	9	8	7
1	5	6	8	9	7	4	2	3

EXPERT - 70

9	7	1	3	2	8	6	5	4
5	6	2	4	9	1	3	8	7
4	8	3	5	6	7	9	2	1
6	1	9	8	7	5	4	3	2
2	5	8	9	3	4	7	1	6
7	3	4	6	1	2	8	9	5
3	4	5	2	8	6	1	7	9
1	9	6	7	5	3	2	4	8
8	2	7	1	4	9	5	6	3

EXPERT - 71

2	5	4	1	6	8	3	9	7
7	6	8	9	3	2	5	1	4
9	3	1	7	5	4	8	6	2
1	9	3	8	7	5	2	4	6
8	4	7	2	1	6	9	5	3
5	2	6	3	4	9	1	7	8
3	8	5	6	9	7	4	2	1
6	1	9	4	2	3	7	8	5
4	7	2	5	8	1	6	3	9

EXPERT - 72

6	8	2	1	9	5	3	4	7
1	7	4	6	8	3	5	2	9
9	5	3	7	2	4	1	8	6
3	2	9	8	6	7	4	1	5
8	4	7	5	3	1	6	9	2
5	1	6	2	4	9	8	7	3
2	3	1	4	7	6	9	5	8
7	9	5	3	1	8	2	6	4
4	6	8	9	5	2	7	3	1

EXPERT - 73

3	7	8	4	6	5	9	2	1
5	6	9	2	1	3	4	8	7
2	4	1	9	8	7	3	6	5
1	8	4	7	2	6	5	3	9
9	5	3	8	4	1	6	7	2
6	2	7	5	3	9	1	4	8
8	3	5	6	9	2	7	1	4
4	9	6	1	7	8	2	5	3
7	1	2	3	5	4	8	9	6

EXPERT - 74

9	6	7	4	2	8	5	3	1
5	8	1	3	9	7	6	4	2
3	4	2	5	6	1	9	8	7
2	9	5	7	8	6	4	1	3
4	7	3	1	5	2	8	9	6
6	1	8	9	4	3	2	7	5
1	5	9	6	7	4	3	2	8
8	3	4	2	1	5	7	6	9
7	2	6	8	3	9	1	5	4

EXPERT - 75

4	9	7	8	1	2	3	5	6
6	1	2	7	3	5	9	8	4
5	8	3	6	4	9	7	1	2
7	4	8	9	6	1	2	3	5
9	5	1	3	2	8	4	6	7
3	2	6	4	5	7	8	9	1
2	7	9	5	8	6	1	4	3
8	3	5	1	7	4	6	2	9
1	6	4	2	9	3	5	7	8

EXPERT - 76

9	8	4	6	1	5	3	7	2
6	3	5	4	2	7	1	9	8
7	1	2	9	8	3	6	4	5
2	5	1	7	9	8	4	6	3
8	7	9	3	6	4	5	2	1
3	4	6	2	5	1	9	8	7
5	6	7	8	3	9	2	1	4
1	2	8	5	4	6	7	3	9
4	9	3	1	7	2	8	5	6

EXPERT - 77

1	3	4	8	6	2	9	5	7
5	7	8	1	4	9	3	6	2
6	9	2	5	3	7	4	8	1
8	2	1	7	9	5	6	3	4
7	4	6	2	8	3	5	1	9
3	5	9	6	1	4	2	7	8
2	1	3	4	5	8	7	9	6
9	8	7	3	2	6	1	4	5
4	6	5	9	7	1	8	2	3

EXPERT - 78

1	5	9	7	8	6	2	3	4
8	7	2	1	3	4	9	6	5
6	4	3	2	9	5	8	7	1
5	2	6	8	4	3	7	1	9
7	8	4	9	2	1	6	5	3
9	3	1	5	6	7	4	8	2
2	6	8	3	1	9	5	4	7
3	9	5	4	7	8	1	2	6
4	1	7	6	5	2	3	9	8

EXPERT - 79

3	4	1	2	5	6	7	8	9
5	6	9	4	8	7	1	2	3
7	2	8	1	9	3	4	5	6
1	8	2	5	4	9	3	6	7
4	7	5	6	3	2	8	9	1
6	9	3	7	1	8	5	4	2
8	1	7	9	2	5	6	3	4
9	5	4	3	6	1	2	7	8
2	3	6	8	7	4	9	1	5

EXPERT - 80

3	8	5	4	1	6	7	9	2
7	9	4	2	5	8	1	6	3
2	6	1	3	7	9	8	5	4
1	3	6	5	9	4	2	8	7
8	4	7	6	2	3	5	1	9
5	2	9	7	8	1	3	4	6
4	1	2	8	6	7	9	3	5
6	7	8	9	3	5	4	2	1
9	5	3	1	4	2	6	7	8

EXPERT - 81

3	9	6	4	5	2	7	8	1
4	8	2	7	3	1	6	9	5
5	7	1	9	8	6	4	2	3
1	2	3	8	6	5	9	4	7
8	4	7	2	1	9	5	3	6
6	5	9	3	7	4	2	1	8
2	6	5	1	9	3	8	7	4
7	1	4	6	2	8	3	5	9
9	3	8	5	4	7	1	6	2

EXPERT - 82

5	9	4	3	2	1	8	7	6
1	3	7	8	6	4	5	2	9
2	8	6	9	7	5	3	4	1
7	4	8	2	1	9	6	5	3
3	1	5	6	4	8	7	9	2
6	2	9	7	5	3	4	1	8
4	5	3	1	8	2	9	6	7
9	6	1	5	3	7	2	8	4
8	7	2	4	9	6	1	3	5

EXPERT - 83

8	2	6	1	7	3	4	5	9
4	5	3	6	2	9	1	8	7
9	7	1	5	8	4	6	3	2
7	1	2	9	4	8	5	6	3
6	3	4	2	1	5	7	9	8
5	9	8	3	6	7	2	4	1
3	4	7	8	5	2	9	1	6
1	8	5	7	9	6	3	2	4
2	6	9	4	3	1	8	7	5

EXPERT - 84

4	5	3	1	6	7	8	9	2
7	1	8	9	2	3	5	4	6
2	9	6	4	5	8	3	7	1
8	6	2	7	1	4	9	3	5
9	3	5	6	8	2	7	1	4
1	7	4	3	9	5	2	6	8
3	2	7	5	4	6	1	8	9
5	4	1	8	7	9	6	2	3
6	8	9	2	3	1	4	5	7

EXPERT - 85

8	6	2	5	7	3	4	9	1
4	9	5	2	6	1	8	7	3
1	7	3	9	8	4	6	2	5
9	3	1	4	2	7	5	8	6
5	2	6	3	9	8	7	1	4
7	8	4	1	5	6	9	3	2
3	5	8	6	1	9	2	4	7
6	4	9	7	3	2	1	5	8
2	1	7	8	4	5	3	6	9

EXPERT - 86

8	7	6	1	5	9	4	2	3
5	4	1	6	3	2	7	8	9
3	2	9	8	7	4	6	5	1
9	5	7	3	4	1	8	6	2
4	1	3	2	8	6	5	9	7
2	6	8	7	9	5	1	3	4
7	3	2	5	1	8	9	4	6
1	8	4	9	6	3	2	7	5
6	9	5	4	2	7	3	1	8

EXPERT - 87

7	8	2	5	1	3	9	6	4
9	3	4	7	6	2	8	1	5
5	6	1	4	9	8	7	3	2
1	7	6	8	4	9	2	5	3
3	4	9	2	5	6	1	7	8
2	5	8	1	3	7	6	4	9
4	1	7	9	8	5	3	2	6
6	9	5	3	2	1	4	8	7
8	2	3	6	7	4	5	9	1

EXPERT - 88

2	9	8	5	3	7	1	6	4
3	4	5	1	6	9	8	2	7
6	7	1	2	4	8	5	3	9
7	3	6	9	2	1	4	8	5
5	1	4	3	8	6	9	7	2
9	8	2	7	5	4	6	1	3
8	5	9	6	7	3	2	4	1
4	2	7	8	1	5	3	9	6
1	6	3	4	9	2	7	5	8

EXPERT - 89

9	6	2	5	7	3	4	1	8
1	4	3	9	8	2	5	7	6
7	8	5	4	6	1	9	3	2
6	7	8	3	4	5	1	2	9
2	3	9	6	1	8	7	5	4
4	5	1	2	9	7	8	6	3
3	1	7	8	2	4	6	9	5
8	2	6	1	5	9	3	4	7
5	9	4	7	3	6	2	8	1

EXPERT - 90

6	3	1	2	5	9	4	8	7
8	7	2	6	4	3	9	1	5
9	4	5	8	7	1	3	2	6
3	8	9	1	6	4	7	5	2
7	2	4	9	8	5	6	3	1
5	1	6	3	2	7	8	4	9
4	9	7	5	3	2	1	6	8
1	5	8	4	9	6	2	7	3
2	6	3	7	1	8	5	9	4

EXPERT - 91

7	9	4	6	2	8	5	3	1
6	5	1	4	7	3	2	8	9
2	8	3	9	1	5	4	7	6
8	2	6	1	3	4	7	9	5
9	1	5	2	8	7	3	6	4
3	4	7	5	6	9	8	1	2
4	6	8	3	9	2	1	5	7
1	7	2	8	5	6	9	4	3
5	3	9	7	4	1	6	2	8

EXPERT - 92

2	4	8	1	3	7	9	5	6
9	1	3	2	6	5	4	7	8
5	7	6	8	9	4	1	3	2
1	9	2	5	4	6	7	8	3
7	8	5	3	1	2	6	9	4
3	6	4	7	8	9	2	1	5
8	2	7	4	5	1	3	6	9
6	5	1	9	2	3	8	4	7
4	3	9	6	7	8	5	2	1

EXPERT - 93

1	9	4	6	5	8	3	2	7
2	8	5	7	3	1	9	6	4
3	6	7	9	2	4	1	8	5
5	4	6	3	7	2	8	1	9
7	3	8	5	1	9	2	4	6
9	2	1	8	4	6	5	7	3
4	5	2	1	9	7	6	3	8
6	1	3	4	8	5	7	9	2
8	7	9	2	6	3	4	5	1

EXPERT - 94

1	5	8	6	3	7	2	4	9
2	9	3	1	8	4	7	5	6
7	4	6	2	5	9	3	1	8
5	2	1	8	7	6	4	9	3
3	7	9	4	2	5	6	8	1
6	8	4	3	9	1	5	7	2
8	3	5	7	1	2	9	6	4
4	1	7	9	6	3	8	2	5
9	6	2	5	4	8	1	3	7

EXPERT - 95

4	6	2	8	5	7	3	9	1
9	5	3	6	4	1	8	7	2
7	8	1	2	3	9	6	5	4
5	1	9	4	6	3	7	2	8
2	4	6	7	9	8	5	1	3
8	3	7	5	1	2	9	4	6
6	9	5	1	8	4	2	3	7
1	2	8	3	7	5	4	6	9
3	7	4	9	2	6	1	8	5

EXPERT - 96

3	9	5	4	7	6	8	1	2
6	8	2	9	1	3	7	5	4
7	4	1	2	8	5	3	9	6
1	2	3	6	9	7	5	4	8
9	5	4	3	2	8	1	6	7
8	6	7	5	4	1	2	3	9
4	3	8	1	6	2	9	7	5
5	7	9	8	3	4	6	2	1
2	1	6	7	5	9	4	8	3

EXPERT - 97

3	2	1	7	5	6	8	9	4
5	9	7	1	4	8	6	2	3
4	6	8	9	3	2	5	7	1
8	1	9	4	7	5	2	3	6
2	5	6	3	8	9	1	4	7
7	3	4	2	6	1	9	8	5
1	8	3	6	2	4	7	5	9
9	4	2	5	1	7	3	6	8
6	7	5	8	9	3	4	1	2

EXPERT - 98

1	7	8	2	6	3	4	5	9
4	3	2	1	5	9	6	8	7
9	5	6	8	4	7	1	2	3
6	8	5	9	2	1	7	3	4
2	9	7	6	3	4	8	1	5
3	4	1	5	7	8	2	9	6
7	2	4	3	8	5	9	6	1
8	1	3	4	9	6	5	7	2
5	6	9	7	1	2	3	4	8

EXPERT - 99

3	2	1	7	8	9	6	5	4
6	5	8	2	1	4	3	7	9
7	4	9	3	5	6	1	2	8
8	3	5	1	7	2	4	9	6
4	7	2	9	6	8	5	1	3
9	1	6	5	4	3	7	8	2
2	8	7	6	3	5	9	4	1
5	9	3	4	2	1	8	6	7
1	6	4	8	9	7	2	3	5

EXPERT - 100

1	5	7	9	6	8	4	3	2
3	4	9	7	2	1	8	5	6
6	8	2	4	3	5	1	7	9
8	9	6	5	1	3	7	2	4
7	3	5	8	4	2	9	6	1
2	1	4	6	7	9	5	8	3
9	7	3	1	5	6	2	4	8
4	2	8	3	9	7	6	1	5
5	6	1	2	8	4	3	9	7

EXPERT - 101

6	9	5	1	2	8	7	4	3
8	7	4	3	5	6	1	2	9
1	2	3	4	7	9	5	6	8
4	8	7	5	9	2	6	3	1
9	1	2	6	8	3	4	5	7
3	5	6	7	1	4	8	9	2
5	4	8	9	3	7	2	1	6
7	6	9	2	4	1	3	8	5
2	3	1	8	6	5	9	7	4

EXPERT - 102

8	4	9	7	2	5	3	1	6
3	7	2	6	8	1	4	9	5
1	6	5	4	9	3	8	7	2
7	8	3	2	1	6	9	5	4
4	2	6	5	7	9	1	3	8
9	5	1	3	4	8	6	2	7
2	1	7	8	3	4	5	6	9
6	9	4	1	5	7	2	8	3
5	3	8	9	6	2	7	4	1

EXPERT - 103

8	3	6	2	1	9	7	5	4
2	9	4	3	7	5	6	8	1
5	1	7	8	4	6	2	3	9
1	2	5	6	3	7	4	9	8
4	6	9	1	8	2	3	7	5
7	8	3	5	9	4	1	2	6
3	5	8	4	2	1	9	6	7
9	4	2	7	6	8	5	1	3
6	7	1	9	5	3	8	4	2

EXPERT - 104

2	1	3	7	6	8	5	9	4
6	9	4	5	1	3	2	7	8
8	7	5	2	4	9	3	1	6
4	3	2	9	5	6	1	8	7
1	8	7	4	3	2	6	5	9
5	6	9	8	7	1	4	3	2
9	2	6	3	8	5	7	4	1
3	4	1	6	9	7	8	2	5
7	5	8	1	2	4	9	6	3

EXPERT - 105

1	8	9	3	7	5	6	2	4
6	5	7	2	4	1	8	3	9
2	3	4	9	6	8	5	7	1
9	4	8	7	5	3	1	6	2
7	2	3	8	1	6	4	9	5
5	1	6	4	2	9	3	8	7
8	6	5	1	9	7	2	4	3
3	9	2	5	8	4	7	1	6
4	7	1	6	3	2	9	5	8

EXPERT - 106

4	1	5	6	9	7	3	2	8
7	6	9	8	3	2	5	4	1
8	2	3	4	5	1	9	7	6
6	5	1	3	4	8	7	9	2
9	7	2	1	6	5	4	8	3
3	4	8	7	2	9	1	6	5
5	3	4	9	8	6	2	1	7
2	8	7	5	1	4	6	3	9
1	9	6	2	7	3	8	5	4

EXPERT - 107

1	9	6	3	2	8	7	4	5
2	7	4	5	1	6	3	9	8
5	3	8	9	7	4	2	1	6
4	8	9	6	5	7	1	2	3
7	2	1	8	9	3	5	6	4
3	6	5	2	4	1	8	7	9
8	5	7	4	6	2	9	3	1
9	4	2	1	3	5	6	8	7
6	1	3	7	8	9	4	5	2

EXPERT - 108

4	5	9	3	7	6	2	1	8
6	2	7	8	1	4	3	5	9
1	8	3	9	2	5	4	7	6
2	7	8	1	5	9	6	4	3
3	6	1	4	8	2	7	9	5
9	4	5	7	6	3	8	2	1
5	1	4	2	3	8	9	6	7
8	9	6	5	4	7	1	3	2
7	3	2	6	9	1	5	8	4

EXPERT - 109

7	6	1	4	9	3	8	2	5
8	4	3	2	1	5	6	9	7
9	2	5	8	6	7	1	3	4
2	9	6	1	7	4	5	8	3
1	3	4	6	5	8	9	7	2
5	7	8	9	3	2	4	1	6
4	5	9	7	2	1	3	6	8
6	8	7	3	4	9	2	5	1
3	1	2	5	8	6	7	4	9

EXPERT - 110

3	2	5	4	9	8	7	1	6
7	9	6	5	1	3	4	2	8
4	1	8	7	6	2	3	9	5
5	4	9	6	8	7	2	3	1
6	3	7	2	5	1	8	4	9
1	8	2	9	3	4	6	5	7
9	6	3	8	4	5	1	7	2
8	7	1	3	2	9	5	6	4
2	5	4	1	7	6	9	8	3

EXPERT - 111

1	9	4	2	8	6	7	5	3
3	8	6	4	7	5	9	1	2
5	7	2	3	9	1	4	6	8
6	3	9	7	4	2	1	8	5
7	2	1	8	5	3	6	9	4
8	4	5	1	6	9	3	2	7
2	5	7	6	1	4	8	3	9
4	6	3	9	2	8	5	7	1
9	1	8	5	3	7	2	4	6

EXPERT - 112

3	7	1	2	5	9	8	6	4
6	2	9	4	8	1	5	3	7
5	4	8	6	7	3	2	1	9
9	1	7	3	6	5	4	8	2
8	3	2	7	1	4	9	5	6
4	5	6	9	2	8	3	7	1
7	9	5	1	3	2	6	4	8
1	8	4	5	9	6	7	2	3
2	6	3	8	4	7	1	9	5

EXPERT - 113

8	7	2	1	4	9	3	5	6
9	3	6	8	2	5	1	4	7
4	1	5	7	6	3	2	9	8
1	6	9	5	7	2	4	8	3
3	5	8	4	1	6	7	2	9
7	2	4	9	3	8	5	6	1
2	9	7	6	5	1	8	3	4
5	8	1	3	9	4	6	7	2
6	4	3	2	8	7	9	1	5

EXPERT - 114

6	4	1	7	9	2	5	3	8
2	7	5	8	4	3	6	9	1
8	3	9	5	1	6	4	2	7
7	9	2	4	3	5	8	1	6
3	8	4	1	6	7	2	5	9
5	1	6	9	2	8	3	7	4
9	5	8	3	7	4	1	6	2
4	2	7	6	5	1	9	8	3
1	6	3	2	8	9	7	4	5

EXPERT - 115

6	5	8	1	7	2	4	3	9
1	7	9	3	6	4	2	8	5
4	2	3	8	5	9	1	6	7
7	4	1	5	3	6	9	2	8
8	6	2	9	4	1	5	7	3
3	9	5	7	2	8	6	1	4
5	3	6	2	9	7	8	4	1
2	8	7	4	1	5	3	9	6
9	1	4	6	8	3	7	5	2

EXPERT - 116

8	1	3	4	5	7	9	6	2
5	2	4	6	9	8	7	1	3
9	7	6	2	3	1	8	4	5
3	5	2	7	8	4	6	9	1
4	8	1	5	6	9	3	2	7
7	6	9	1	2	3	4	5	8
2	4	5	8	7	6	1	3	9
6	9	8	3	1	5	2	7	4
1	3	7	9	4	2	5	8	6

EXPERT - 117

7	2	1	8	9	4	5	3	6
9	6	5	7	3	1	8	4	2
4	3	8	2	6	5	1	7	9
2	8	3	1	5	6	4	9	7
5	9	7	4	2	3	6	1	8
1	4	6	9	8	7	3	2	5
8	1	4	6	7	9	2	5	3
6	5	9	3	1	2	7	8	4
3	7	2	5	4	8	9	6	1

EXPERT - 118

2	3	1	4	9	5	6	8	7
4	6	7	8	3	1	9	2	5
5	8	9	2	6	7	4	1	3
7	2	8	6	4	9	3	5	1
6	9	3	5	1	2	7	4	8
1	5	4	7	8	3	2	9	6
8	7	6	9	5	4	1	3	2
9	1	5	3	2	6	8	7	4
3	4	2	1	7	8	5	6	9

EXPERT - 119

2	4	9	6	7	1	8	5	3
1	8	7	3	5	9	6	4	2
6	5	3	4	2	8	7	1	9
9	6	4	8	3	2	1	7	5
7	1	8	5	9	4	2	3	6
3	2	5	7	1	6	4	9	8
8	3	1	2	4	5	9	6	7
4	7	2	9	6	3	5	8	1
5	9	6	1	8	7	3	2	4

EXPERT - 120

7	2	4	6	8	1	3	5	9
5	1	8	3	9	7	2	6	4
3	9	6	4	5	2	7	1	8
1	7	9	5	3	4	8	2	6
6	5	2	8	1	9	4	3	7
4	8	3	7	2	6	1	9	5
9	4	1	2	6	8	5	7	3
8	6	5	1	7	3	9	4	2
2	3	7	9	4	5	6	8	1

EXPERT - 121

7	9	6	3	2	1	5	8	4
2	5	3	7	4	8	9	6	1
1	8	4	5	6	9	7	2	3
4	3	5	1	7	6	8	9	2
8	2	1	9	3	5	6	4	7
9	6	7	4	8	2	1	3	5
5	7	8	2	9	3	4	1	6
3	4	9	6	1	7	2	5	8
6	1	2	8	5	4	3	7	9

EXPERT - 122

8	6	5	4	1	3	7	9	2
4	9	7	6	8	2	5	1	3
2	3	1	5	9	7	4	8	6
5	1	3	9	6	4	8	2	7
7	4	9	2	5	8	6	3	1
6	8	2	7	3	1	9	4	5
9	2	8	1	7	5	3	6	4
1	5	6	3	4	9	2	7	8
3	7	4	8	2	6	1	5	9

EXPERT - 123

4	9	6	3	5	1	2	8	7
3	8	1	2	6	7	9	4	5
5	7	2	8	9	4	3	1	6
9	2	3	5	1	6	8	7	4
6	5	7	4	3	8	1	2	9
8	1	4	9	7	2	5	6	3
1	4	5	7	8	9	6	3	2
7	3	8	6	2	5	4	9	1
2	6	9	1	4	3	7	5	8

EXPERT - 124

5	4	9	6	2	3	1	7	8
6	2	1	7	8	5	9	4	3
3	7	8	1	9	4	2	5	6
8	1	4	9	3	7	6	2	5
9	5	2	8	1	6	4	3	7
7	6	3	5	4	2	8	9	1
2	3	5	4	6	8	7	1	9
1	8	7	2	5	9	3	6	4
4	9	6	3	7	1	5	8	2

EXPERT - 125

6	8	3	2	5	7	4	1	9
7	4	5	9	3	1	2	6	8
1	2	9	8	4	6	3	7	5
5	7	4	1	6	3	9	8	2
3	6	8	4	2	9	7	5	1
2	9	1	7	8	5	6	3	4
4	1	2	6	7	8	5	9	3
9	3	7	5	1	2	8	4	6
8	5	6	3	9	4	1	2	7

EXPERT - 126

3	5	2	1	8	6	7	4	9
8	4	6	7	9	5	2	3	1
1	9	7	2	3	4	8	5	6
5	6	3	4	7	1	9	8	2
2	7	9	6	5	8	4	1	3
4	1	8	9	2	3	5	6	7
6	2	5	3	4	7	1	9	8
9	8	1	5	6	2	3	7	4
7	3	4	8	1	9	6	2	5

EXPERT - 127

5	6	3	9	8	4	1	2	7
1	4	9	2	7	6	3	5	8
8	7	2	5	3	1	4	6	9
2	8	5	4	6	9	7	3	1
9	3	6	1	2	7	5	8	4
4	1	7	8	5	3	2	9	6
6	9	1	3	4	5	8	7	2
7	5	8	6	1	2	9	4	3
3	2	4	7	9	8	6	1	5

EXPERT - 128

7	9	5	8	3	2	1	4	6
4	8	3	1	6	7	2	5	9
2	1	6	5	4	9	3	7	8
8	5	9	2	1	3	4	6	7
1	7	4	9	5	6	8	2	3
3	6	2	4	7	8	9	1	5
5	3	1	7	8	4	6	9	2
6	2	7	3	9	1	5	8	4
9	4	8	6	2	5	7	3	1

EXPERT - 129

4	8	9	5	2	1	6	7	3
6	1	3	8	4	7	2	5	9
5	2	7	9	3	6	1	8	4
2	7	1	3	8	9	5	4	6
8	6	5	1	7	4	3	9	2
9	3	4	2	6	5	7	1	8
1	4	6	7	9	3	8	2	5
3	5	8	4	1	2	9	6	7
7	9	2	6	5	8	4	3	1

EXPERT - 130

3	2	5	1	7	9	8	4	6
9	6	1	4	2	8	3	7	5
7	4	8	6	5	3	1	2	9
2	7	3	9	4	5	6	1	8
4	1	9	8	6	2	5	3	7
8	5	6	7	3	1	2	9	4
5	9	2	3	8	7	4	6	1
6	3	7	5	1	4	9	8	2
1	8	4	2	9	6	7	5	3

EXPERT - 131

9	3	5	4	6	2	7	8	1
6	2	4	7	8	1	9	3	5
1	8	7	9	3	5	4	2	6
8	1	2	3	4	9	5	6	7
7	9	6	5	2	8	3	1	4
5	4	3	1	7	6	2	9	8
2	6	9	8	5	7	1	4	3
4	7	1	6	9	3	8	5	2
3	5	8	2	1	4	6	7	9

EXPERT - 132

5	1	9	4	7	3	2	8	6
2	3	7	9	6	8	1	4	5
4	8	6	5	1	2	3	9	7
9	6	1	8	4	7	5	2	3
7	5	8	3	2	9	4	6	1
3	4	2	6	5	1	8	7	9
1	9	4	7	8	5	6	3	2
8	7	5	2	3	6	9	1	4
6	2	3	1	9	4	7	5	8

EXPERT - 133

1	7	3	2	4	8	6	9	5
8	9	4	3	6	5	2	1	7
6	5	2	7	9	1	4	3	8
5	4	6	9	8	3	7	2	1
2	3	9	5	1	7	8	4	6
7	8	1	4	2	6	9	5	3
4	6	8	1	5	9	3	7	2
3	2	5	6	7	4	1	8	9
9	1	7	8	3	2	5	6	4

EXPERT - 134

9	1	6	5	4	7	2	8	3
7	8	2	3	6	1	4	5	9
5	4	3	9	2	8	6	7	1
4	2	8	1	3	5	7	9	6
3	9	1	8	7	6	5	2	4
6	5	7	4	9	2	3	1	8
2	6	9	7	1	4	8	3	5
1	7	5	6	8	3	9	4	2
8	3	4	2	5	9	1	6	7

EXPERT - 135

3	2	8	9	6	1	5	4	7
5	1	4	7	3	8	9	6	2
9	7	6	4	5	2	1	8	3
4	9	5	2	1	3	6	7	8
6	3	1	8	7	4	2	9	5
2	8	7	5	9	6	3	1	4
8	5	9	6	2	7	4	3	1
7	6	3	1	4	5	8	2	9
1	4	2	3	8	9	7	5	6

EXPERT - 136

9	7	3	6	5	4	8	2	1
4	6	2	1	9	8	7	5	3
5	1	8	2	3	7	4	6	9
6	9	7	3	4	1	2	8	5
1	3	4	5	8	2	6	9	7
8	2	5	7	6	9	1	3	4
3	8	9	4	1	6	5	7	2
7	5	1	8	2	3	9	4	6
2	4	6	9	7	5	3	1	8

EXPERT - 137

3	9	5	7	2	1	8	6	4
1	8	6	9	3	4	2	7	5
4	2	7	8	6	5	1	9	3
8	1	2	4	7	3	6	5	9
9	6	4	5	8	2	3	1	7
7	5	3	6	1	9	4	8	2
2	3	8	1	9	7	5	4	6
6	4	9	2	5	8	7	3	1
5	7	1	3	4	6	9	2	8

EXPERT - 138

6	4	2	8	5	9	7	1	3
8	5	3	1	7	2	6	4	9
7	1	9	6	3	4	5	2	8
5	8	6	3	2	1	4	9	7
2	9	4	7	8	5	1	3	6
1	3	7	9	4	6	2	8	5
9	7	5	4	1	3	8	6	2
4	6	8	2	9	7	3	5	1
3	2	1	5	6	8	9	7	4

EXPERT - 139

6	5	2	3	9	7	1	4	8
9	8	4	5	1	6	7	3	2
1	3	7	2	8	4	9	6	5
8	1	6	4	2	5	3	9	7
3	2	5	6	7	9	8	1	4
7	4	9	1	3	8	5	2	6
4	9	3	8	5	2	6	7	1
5	6	1	7	4	3	2	8	9
2	7	8	9	6	1	4	5	3

EXPERT - 140

7	6	8	1	9	2	4	5	3
4	9	5	8	3	6	2	1	7
1	2	3	7	5	4	8	6	9
8	1	9	6	2	7	3	4	5
2	5	4	3	1	9	7	8	6
6	3	7	4	8	5	9	2	1
5	8	2	9	6	3	1	7	4
3	4	6	2	7	1	5	9	8
9	7	1	5	4	8	6	3	2

EXPERT - 141

8	4	9	1	7	6	2	5	3
2	7	3	5	9	4	8	1	6
1	6	5	2	3	8	9	7	4
7	8	2	6	4	5	3	9	1
9	3	6	7	2	1	4	8	5
5	1	4	9	8	3	6	2	7
4	2	8	3	5	7	1	6	9
3	5	1	8	6	9	7	4	2
6	9	7	4	1	2	5	3	8

EXPERT - 142

5	9	3	1	8	7	6	2	4
4	2	7	5	6	3	9	8	1
1	6	8	2	9	4	5	7	3
7	8	4	3	1	6	2	5	9
9	5	1	4	2	8	7	3	6
6	3	2	7	5	9	4	1	8
2	7	9	8	4	1	3	6	5
8	4	5	6	3	2	1	9	7
3	1	6	9	7	5	8	4	2

EXPERT - 143

9	2	6	1	3	8	4	5	7
1	5	4	2	7	6	3	9	8
3	7	8	9	4	5	2	1	6
2	6	1	5	8	4	7	3	9
8	3	9	6	1	7	5	2	4
5	4	7	3	9	2	6	8	1
4	1	2	8	6	3	9	7	5
7	9	3	4	5	1	8	6	2
6	8	5	7	2	9	1	4	3

EXPERT - 144

5	9	1	6	8	4	3	7	2
7	6	2	3	5	1	4	8	9
8	4	3	7	9	2	5	1	6
9	5	8	2	6	7	1	3	4
6	2	4	1	3	8	7	9	5
1	3	7	9	4	5	6	2	8
2	8	5	4	7	3	9	6	1
3	1	6	5	2	9	8	4	7
4	7	9	8	1	6	2	5	3

EXPERT - 145

4	9	5	6	1	2	3	8	7
6	7	3	5	4	8	9	2	1
1	2	8	7	3	9	4	5	6
8	5	6	3	9	1	7	4	2
2	1	4	8	7	5	6	3	9
7	3	9	4	2	6	5	1	8
9	8	7	1	5	3	2	6	4
3	6	2	9	8	4	1	7	5
5	4	1	2	6	7	8	9	3

EXPERT - 146

3	9	6	5	1	2	8	7	4
1	7	4	6	9	8	3	2	5
2	5	8	3	7	4	6	1	9
4	1	5	9	8	6	7	3	2
7	2	3	4	5	1	9	8	6
8	6	9	7	2	3	5	4	1
6	4	2	8	3	5	1	9	7
5	3	7	1	4	9	2	6	8
9	8	1	2	6	7	4	5	3

EXPERT - 147

4	8	3	2	5	1	7	6	9
1	5	9	7	8	6	3	2	4
2	7	6	3	9	4	8	5	1
6	4	7	9	2	3	5	1	8
3	1	5	4	6	8	2	9	7
9	2	8	1	7	5	6	4	3
5	9	4	6	3	7	1	8	2
8	3	2	5	1	9	4	7	6
7	6	1	8	4	2	9	3	5

EXPERT - 148

5	7	4	9	1	2	6	3	8
9	8	6	7	3	5	1	4	2
3	1	2	4	8	6	7	5	9
1	2	9	3	6	4	5	8	7
7	6	8	2	5	9	3	1	4
4	3	5	8	7	1	2	9	6
6	5	7	1	9	8	4	2	3
8	4	3	5	2	7	9	6	1
2	9	1	6	4	3	8	7	5

EXPERT - 149

8	2	6	4	9	1	5	7	3
3	4	1	7	5	6	9	2	8
7	5	9	8	3	2	6	4	1
5	3	2	6	1	7	4	8	9
9	7	4	5	2	8	3	1	6
1	6	8	3	4	9	7	5	2
2	9	3	1	7	5	8	6	4
6	1	7	9	8	4	2	3	5
4	8	5	2	6	3	1	9	7

EXPERT - 150

2	3	9	4	8	1	6	5	7
7	1	6	9	5	2	8	4	3
8	5	4	7	6	3	9	2	1
9	4	2	3	1	5	7	8	6
1	8	7	6	9	4	5	3	2
5	6	3	8	2	7	1	9	4
4	9	8	2	7	6	3	1	5
6	2	5	1	3	8	4	7	9
3	7	1	5	4	9	2	6	8

EXPERT - 151

1	7	6	4	5	8	2	3	9
3	4	9	6	2	1	5	7	8
5	2	8	7	3	9	6	1	4
6	3	2	5	9	7	4	8	1
9	8	4	3	1	2	7	6	5
7	5	1	8	4	6	3	9	2
4	1	7	9	6	5	8	2	3
8	9	3	2	7	4	1	5	6
2	6	5	1	8	3	9	4	7

EXPERT - 152

6	8	4	9	2	7	5	3	1
3	5	9	1	4	6	2	8	7
7	2	1	8	3	5	4	6	9
1	4	8	5	7	3	9	2	6
5	6	3	2	9	1	7	4	8
2	9	7	4	6	8	1	5	3
4	7	6	3	5	9	8	1	2
8	3	5	7	1	2	6	9	4
9	1	2	6	8	4	3	7	5

EXPERT - 153

6	1	5	2	7	9	4	3	8
8	2	9	1	4	3	7	5	6
7	3	4	5	6	8	9	2	1
2	5	3	4	8	7	1	6	9
4	9	7	6	1	5	3	8	2
1	8	6	9	3	2	5	4	7
5	6	8	3	9	1	2	7	4
3	7	1	8	2	4	6	9	5
9	4	2	7	5	6	8	1	3

EXPERT - 154

8	9	6	1	5	4	3	2	7
4	7	5	9	2	3	1	8	6
1	3	2	6	7	8	4	5	9
6	2	8	7	4	9	5	1	3
5	4	3	8	6	1	7	9	2
7	1	9	5	3	2	6	4	8
2	8	7	4	1	6	9	3	5
9	6	4	3	8	5	2	7	1
3	5	1	2	9	7	8	6	4

EXPERT - 155

9	4	7	1	6	8	3	2	5
1	6	2	4	5	3	9	7	8
5	8	3	2	7	9	4	1	6
7	9	6	3	2	4	8	5	1
2	5	4	6	8	1	7	3	9
8	3	1	5	9	7	6	4	2
6	2	9	7	4	5	1	8	3
4	1	8	9	3	2	5	6	7
3	7	5	8	1	6	2	9	4

EXPERT - 156

4	1	2	7	3	5	8	6	9
7	8	5	9	2	6	1	3	4
9	6	3	4	1	8	7	5	2
3	2	1	6	7	4	9	8	5
5	4	9	1	8	3	6	2	7
8	7	6	2	5	9	4	1	3
1	9	8	5	4	2	3	7	6
6	5	7	3	9	1	2	4	8
2	3	4	8	6	7	5	9	1

EXPERT - 157

2	6	3	5	1	7	9	8	4
1	9	4	3	8	2	5	6	7
5	7	8	9	6	4	2	3	1
8	3	7	1	9	6	4	5	2
9	2	6	4	5	3	7	1	8
4	1	5	2	7	8	3	9	6
3	8	2	6	4	5	1	7	9
7	5	9	8	2	1	6	4	3
6	4	1	7	3	9	8	2	5

EXPERT - 158

5	2	9	6	8	1	4	3	7
8	7	3	5	9	4	6	2	1
1	4	6	3	2	7	5	9	8
2	6	1	8	7	3	9	4	5
9	5	8	2	4	6	7	1	3
4	3	7	9	1	5	2	8	6
3	8	2	7	6	9	1	5	4
6	1	5	4	3	2	8	7	9
7	9	4	1	5	8	3	6	2

EXPERT - 159

9	6	4	8	5	2	7	3	1
7	8	2	1	3	9	5	6	4
1	3	5	7	4	6	2	8	9
6	7	3	5	8	1	9	4	2
2	4	9	3	6	7	1	5	8
5	1	8	2	9	4	6	7	3
3	9	7	4	1	5	8	2	6
4	5	6	9	2	8	3	1	7
8	2	1	6	7	3	4	9	5

EXPERT - 160

7	9	2	8	5	4	6	1	3
4	1	6	7	2	3	8	9	5
8	3	5	1	9	6	2	4	7
9	7	1	4	6	2	3	5	8
5	2	4	3	7	8	1	6	9
6	8	3	9	1	5	7	2	4
1	6	8	5	4	7	9	3	2
3	5	9	2	8	1	4	7	6
2	4	7	6	3	9	5	8	1

EXPERT - 161

4	2	5	6	1	7	8	9	3
9	7	3	4	2	8	6	1	5
1	6	8	9	3	5	2	7	4
7	3	4	2	6	9	5	8	1
8	5	2	1	7	3	9	4	6
6	9	1	5	8	4	3	2	7
2	8	7	3	5	1	4	6	9
3	1	9	8	4	6	7	5	2
5	4	6	7	9	2	1	3	8

EXPERT - 162

9	5	8	3	2	7	1	4	6
6	4	7	8	5	1	9	2	3
2	1	3	6	4	9	5	8	7
8	9	4	1	3	5	7	6	2
7	6	2	9	8	4	3	1	5
1	3	5	7	6	2	8	9	4
3	8	9	2	7	6	4	5	1
5	2	1	4	9	3	6	7	8
4	7	6	5	1	8	2	3	9

EXPERT - 163

2	5	8	3	9	1	4	7	6
4	1	7	6	8	2	5	9	3
6	3	9	7	4	5	8	2	1
8	6	1	9	2	4	3	5	7
7	9	3	1	5	6	2	8	4
5	4	2	8	3	7	1	6	9
1	8	6	2	7	3	9	4	5
3	2	5	4	6	9	7	1	8
9	7	4	5	1	8	6	3	2

EXPERT - 164

2	3	5	4	7	1	6	8	9
4	6	9	2	8	5	7	1	3
7	1	8	3	6	9	2	5	4
5	9	3	6	2	4	8	7	1
8	7	2	1	9	3	4	6	5
1	4	6	7	5	8	3	9	2
9	5	7	8	4	2	1	3	6
6	2	1	5	3	7	9	4	8
3	8	4	9	1	6	5	2	7

EXPERT - 165

4	5	7	2	1	9	8	6	3
6	2	9	4	3	8	5	7	1
8	1	3	5	7	6	2	9	4
7	4	5	9	2	1	6	3	8
9	3	6	8	4	5	1	2	7
2	8	1	7	6	3	4	5	9
1	7	8	3	5	2	9	4	6
5	9	4	6	8	7	3	1	2
3	6	2	1	9	4	7	8	5

EXPERT - 166

4	7	1	3	8	2	6	9	5
9	2	5	1	6	7	4	8	3
8	6	3	5	4	9	1	7	2
7	5	4	6	1	8	2	3	9
1	8	6	2	9	3	7	5	4
3	9	2	4	7	5	8	6	1
2	3	7	8	5	4	9	1	6
6	4	8	9	3	1	5	2	7
5	1	9	7	2	6	3	4	8

EXPERT - 167

6	2	5	9	8	7	3	4	1
7	4	3	1	6	2	9	8	5
9	8	1	3	5	4	7	2	6
8	7	2	6	3	1	5	9	4
4	3	6	5	2	9	1	7	8
1	5	9	4	7	8	6	3	2
3	9	8	2	1	6	4	5	7
5	6	7	8	4	3	2	1	9
2	1	4	7	9	5	8	6	3

EXPERT - 168

2	1	5	6	4	3	9	8	7
8	6	7	2	1	9	3	5	4
4	3	9	5	7	8	1	2	6
6	5	2	3	9	1	4	7	8
1	4	3	8	6	7	5	9	2
9	7	8	4	5	2	6	1	3
5	2	6	1	8	4	7	3	9
7	8	1	9	3	6	2	4	5
3	9	4	7	2	5	8	6	1

EXPERT - 169

3	2	1	5	6	8	4	7	9
6	8	7	9	4	1	5	3	2
4	5	9	3	7	2	6	8	1
5	3	2	7	1	4	9	6	8
7	9	4	8	3	6	1	2	5
8	1	6	2	5	9	3	4	7
1	7	8	4	9	3	2	5	6
2	6	3	1	8	5	7	9	4
9	4	5	6	2	7	8	1	3

EXPERT - 170

1	6	2	4	9	8	3	5	7
5	4	9	2	7	3	8	1	6
7	3	8	5	6	1	9	2	4
6	8	1	7	5	9	2	4	3
9	5	7	3	2	4	1	6	8
4	2	3	1	8	6	7	9	5
3	7	5	6	1	2	4	8	9
8	1	4	9	3	5	6	7	2
2	9	6	8	4	7	5	3	1

EXPERT - 171

3	7	9	6	2	1	8	4	5
6	8	5	7	4	3	1	9	2
1	4	2	5	9	8	7	3	6
4	5	1	8	3	6	9	2	7
9	3	7	1	5	2	4	6	8
8	2	6	9	7	4	5	1	3
2	1	3	4	8	5	6	7	9
5	9	4	2	6	7	3	8	1
7	6	8	3	1	9	2	5	4

EXPERT - 172

4	2	5	8	1	7	3	9	6
6	1	7	9	3	4	2	5	8
8	3	9	6	2	5	7	1	4
5	6	1	4	7	3	8	2	9
3	9	4	2	5	8	1	6	7
2	7	8	1	9	6	4	3	5
1	8	3	7	6	9	5	4	2
9	4	2	5	8	1	6	7	3
7	5	6	3	4	2	9	8	1

EXPERT - 173

7	8	9	2	6	3	5	4	1
4	6	5	9	1	8	3	7	2
3	1	2	7	5	4	9	8	6
5	2	4	3	8	9	6	1	7
1	9	8	4	7	6	2	3	5
6	3	7	1	2	5	4	9	8
2	7	3	6	9	1	8	5	4
8	4	6	5	3	7	1	2	9
9	5	1	8	4	2	7	6	3

EXPERT - 174

5	4	9	8	1	2	6	3	7
2	6	7	3	4	5	9	1	8
1	3	8	7	9	6	5	4	2
9	7	3	4	2	8	1	6	5
4	8	5	1	6	7	2	9	3
6	2	1	9	5	3	7	8	4
3	5	6	2	8	1	4	7	9
7	9	2	6	3	4	8	5	1
8	1	4	5	7	9	3	2	6

EXPERT - 175

9	4	2	8	3	1	7	5	6
1	7	8	4	5	6	9	3	2
5	3	6	7	2	9	4	1	8
2	6	7	1	9	5	3	8	4
8	9	5	2	4	3	1	6	7
4	1	3	6	8	7	5	2	9
6	5	9	3	7	2	8	4	1
7	2	4	5	1	8	6	9	3
3	8	1	9	6	4	2	7	5

EXPERT - 176

9	2	4	6	1	3	7	5	8
7	1	6	8	2	5	3	9	4
5	3	8	4	9	7	2	6	1
2	4	5	9	3	1	8	7	6
1	8	9	7	6	4	5	3	2
3	6	7	5	8	2	4	1	9
4	7	1	2	5	9	6	8	3
6	5	3	1	4	8	9	2	7
8	9	2	3	7	6	1	4	5

EXPERT - 177

7	4	9	5	1	6	2	8	3
3	8	1	2	4	7	6	9	5
6	5	2	8	9	3	7	4	1
9	1	5	7	2	8	3	6	4
2	7	3	4	6	5	8	1	9
8	6	4	1	3	9	5	2	7
5	9	8	6	7	4	1	3	2
4	2	6	3	5	1	9	7	8
1	3	7	9	8	2	4	5	6

EXPERT - 178

2	1	9	4	3	6	8	7	5
5	4	3	8	7	2	9	1	6
8	7	6	1	5	9	4	3	2
9	8	7	6	1	5	3	2	4
6	3	4	9	2	7	5	8	1
1	2	5	3	8	4	7	6	9
4	6	2	7	9	3	1	5	8
7	9	8	5	6	1	2	4	3
3	5	1	2	4	8	6	9	7

EXPERT - 179

1	5	6	2	3	7	4	9	8
2	9	8	5	1	4	7	3	6
4	3	7	9	6	8	5	1	2
6	4	5	8	9	1	2	7	3
8	7	9	4	2	3	6	5	1
3	1	2	7	5	6	9	8	4
7	8	1	6	4	5	3	2	9
5	2	4	3	8	9	1	6	7
9	6	3	1	7	2	8	4	5

EXPERT - 180

2	6	1	5	3	8	9	4	7
3	8	5	9	7	4	1	2	6
7	9	4	1	2	6	8	5	3
9	7	8	3	4	5	6	1	2
6	4	2	8	1	7	5	3	9
1	5	3	6	9	2	4	7	8
4	3	6	7	8	1	2	9	5
8	1	7	2	5	9	3	6	4
5	2	9	4	6	3	7	8	1

EXPERT - 181

3	8	4	2	7	5	6	1	9
5	2	1	4	6	9	7	3	8
7	9	6	1	3	8	5	2	4
8	7	3	6	1	4	2	9	5
1	4	2	5	9	3	8	7	6
9	6	5	8	2	7	3	4	1
4	3	7	9	8	6	1	5	2
6	1	9	7	5	2	4	8	3
2	5	8	3	4	1	9	6	7

EXPERT - 182

2	5	9	8	1	7	3	6	4
1	3	6	5	4	9	7	2	8
8	7	4	3	6	2	9	1	5
5	6	2	1	9	3	4	8	7
7	9	1	4	8	6	2	5	3
3	4	8	7	2	5	1	9	6
4	2	3	9	5	8	6	7	1
6	8	7	2	3	1	5	4	9
9	1	5	6	7	4	8	3	2

EXPERT - 183

2	6	9	8	4	1	7	3	5
3	8	5	9	2	7	6	4	1
4	1	7	5	6	3	2	9	8
7	5	1	4	3	8	9	2	6
6	3	4	1	9	2	5	8	7
9	2	8	6	7	5	4	1	3
8	4	2	7	1	6	3	5	9
1	9	6	3	5	4	8	7	2
5	7	3	2	8	9	1	6	4

EXPERT - 184

6	9	8	5	2	7	4	1	3
7	2	4	1	3	9	5	8	6
1	5	3	4	6	8	2	9	7
5	1	7	8	9	3	6	4	2
3	6	2	7	1	4	9	5	8
4	8	9	6	5	2	7	3	1
8	4	1	2	7	5	3	6	9
9	7	5	3	8	6	1	2	4
2	3	6	9	4	1	8	7	5

EXPERT - 185

5	7	6	2	8	9	4	1	3
1	8	2	3	4	6	7	5	9
9	3	4	7	1	5	8	6	2
2	1	9	5	7	8	6	3	4
3	6	8	9	2	4	5	7	1
4	5	7	1	6	3	2	9	8
7	4	1	6	3	2	9	8	5
8	9	3	4	5	7	1	2	6
6	2	5	8	9	1	3	4	7

EXPERT - 186

7	6	3	9	5	4	8	1	2
5	1	8	2	6	7	9	3	4
9	4	2	8	1	3	7	6	5
2	3	5	6	4	9	1	7	8
8	7	6	1	2	5	4	9	3
4	9	1	7	3	8	5	2	6
3	8	7	4	9	2	6	5	1
1	2	9	5	8	6	3	4	7
6	5	4	3	7	1	2	8	9

EXPERT - 187

5	9	2	4	6	1	7	8	3
6	4	7	8	9	3	1	2	5
1	8	3	2	5	7	9	4	6
3	5	9	6	1	2	4	7	8
8	7	6	3	4	9	2	5	1
4	2	1	5	7	8	6	3	9
2	6	5	1	8	4	3	9	7
9	3	8	7	2	6	5	1	4
7	1	4	9	3	5	8	6	2

EXPERT - 188

7	9	5	3	4	8	1	6	2
2	3	6	9	1	5	8	7	4
4	8	1	2	6	7	5	3	9
9	4	7	5	8	6	3	2	1
5	6	3	1	2	4	9	8	7
8	1	2	7	9	3	6	4	5
6	2	8	4	5	9	7	1	3
3	5	4	6	7	1	2	9	8
1	7	9	8	3	2	4	5	6

EXPERT - 189

3	7	6	9	4	8	5	1	2
4	5	8	1	6	2	3	9	7
1	2	9	5	7	3	4	6	8
2	6	3	4	8	5	9	7	1
7	4	1	3	2	9	6	8	5
9	8	5	6	1	7	2	3	4
8	9	4	7	5	6	1	2	3
6	1	2	8	3	4	7	5	9
5	3	7	2	9	1	8	4	6

EXPERT - 190

4	3	1	8	5	2	7	9	6
5	9	7	3	1	6	4	2	8
6	2	8	7	4	9	5	1	3
2	5	6	9	7	4	8	3	1
8	4	9	1	2	3	6	7	5
1	7	3	6	8	5	9	4	2
7	1	4	5	3	8	2	6	9
9	8	2	4	6	1	3	5	7
3	6	5	2	9	7	1	8	4

EXPERT - 191

8	9	3	4	5	6	2	1	7
4	2	5	7	1	8	3	9	6
7	1	6	9	3	2	5	8	4
1	3	8	5	4	7	9	6	2
5	7	2	6	8	9	4	3	1
6	4	9	1	2	3	8	7	5
2	5	7	3	9	1	6	4	8
3	6	4	8	7	5	1	2	9
9	8	1	2	6	4	7	5	3

EXPERT - 192

9	6	1	8	7	2	4	3	5
8	7	2	5	4	3	1	6	9
3	4	5	6	9	1	2	8	7
4	2	9	7	3	8	6	5	1
6	1	3	4	5	9	8	7	2
7	5	8	1	2	6	3	9	4
2	9	4	3	8	5	7	1	6
5	3	6	2	1	7	9	4	8
1	8	7	9	6	4	5	2	3

EXPERT - 193

7	2	4	9	6	3	5	8	1
1	6	8	5	7	4	2	9	3
5	3	9	2	8	1	7	6	4
9	1	5	7	4	8	3	2	6
4	8	3	1	2	6	9	7	5
6	7	2	3	5	9	4	1	8
8	5	1	4	9	2	6	3	7
2	4	6	8	3	7	1	5	9
3	9	7	6	1	5	8	4	2

EXPERT - 194

4	6	5	9	7	3	1	8	2
1	2	3	4	5	8	6	7	9
7	8	9	6	1	2	5	4	3
5	9	6	8	4	7	3	2	1
3	4	2	5	6	1	7	9	8
8	7	1	2	3	9	4	5	6
2	5	7	1	9	6	8	3	4
6	3	8	7	2	4	9	1	5
9	1	4	3	8	5	2	6	7

EXPERT - 195

8	4	2	9	6	7	1	3	5
6	9	3	1	5	8	2	7	4
1	5	7	2	3	4	9	8	6
3	6	4	5	7	9	8	2	1
2	7	5	8	1	3	4	6	9
9	1	8	4	2	6	7	5	3
7	3	1	6	4	2	5	9	8
4	2	9	3	8	5	6	1	7
5	8	6	7	9	1	3	4	2

EXPERT - 196

6	4	3	2	7	1	8	5	9
5	9	7	3	8	6	1	2	4
2	8	1	5	4	9	6	7	3
7	1	2	9	6	8	4	3	5
8	3	9	4	5	2	7	1	6
4	6	5	1	3	7	2	9	8
3	2	6	8	1	5	9	4	7
9	7	4	6	2	3	5	8	1
1	5	8	7	9	4	3	6	2

EXPERT - 197

9	4	6	2	8	5	7	1	3
7	8	5	3	6	1	9	2	4
1	3	2	4	9	7	8	5	6
6	9	1	5	4	8	2	3	7
5	7	3	9	1	2	4	6	8
4	2	8	7	3	6	5	9	1
8	1	7	6	2	9	3	4	5
2	5	4	1	7	3	6	8	9
3	6	9	8	5	4	1	7	2

EXPERT - 198

4	8	1	6	9	2	7	3	5
7	2	6	3	5	4	9	8	1
9	3	5	1	7	8	6	2	4
1	7	2	4	3	5	8	6	9
5	9	4	8	6	7	3	1	2
3	6	8	2	1	9	4	5	7
6	4	9	5	2	3	1	7	8
8	5	3	7	4	1	2	9	6
2	1	7	9	8	6	5	4	3

EXPERT - 199

8	9	3	6	7	4	5	1	2
4	2	1	9	5	3	8	7	6
6	5	7	8	2	1	4	9	3
9	6	5	1	4	2	3	8	7
1	8	4	7	3	9	2	6	5
7	3	2	5	8	6	9	4	1
5	1	6	2	9	8	7	3	4
2	4	9	3	6	7	1	5	8
3	7	8	4	1	5	6	2	9

EXPERT - 200

7	9	4	8	3	2	6	1	5
8	5	2	6	1	4	9	3	7
1	6	3	9	5	7	4	2	8
9	3	8	4	7	6	2	5	1
5	4	7	2	8	1	3	6	9
2	1	6	3	9	5	7	8	4
4	2	9	5	6	8	1	7	3
6	7	5	1	4	3	8	9	2
3	8	1	7	2	9	5	4	6

EXPERT - 201

4	6	5	7	9	8	3	2	1
8	2	3	5	1	4	9	6	7
7	1	9	6	2	3	4	5	8
1	4	7	9	8	6	2	3	5
9	8	6	3	5	2	7	1	4
3	5	2	4	7	1	6	8	9
2	7	4	8	6	5	1	9	3
5	3	1	2	4	9	8	7	6
6	9	8	1	3	7	5	4	2

EXPERT - 202

7	6	3	2	4	1	9	5	8
9	4	8	5	7	6	3	1	2
2	5	1	3	8	9	4	6	7
5	9	7	4	3	2	1	8	6
8	3	6	7	1	5	2	4	9
4	1	2	9	6	8	5	7	3
1	7	5	8	9	3	6	2	4
6	8	9	1	2	4	7	3	5
3	2	4	6	5	7	8	9	1

EXPERT - 203

2	5	3	1	7	9	6	4	8
9	1	6	8	3	4	7	2	5
4	8	7	5	6	2	1	3	9
6	4	1	3	9	8	2	5	7
8	2	5	7	1	6	4	9	3
7	3	9	2	4	5	8	1	6
3	7	4	6	5	1	9	8	2
5	9	8	4	2	7	3	6	1
1	6	2	9	8	3	5	7	4

EXPERT - 204

2	5	1	7	6	8	4	9	3
7	4	9	3	1	5	8	2	6
6	3	8	2	9	4	7	1	5
5	8	7	1	4	2	6	3	9
4	9	3	6	5	7	1	8	2
1	6	2	9	8	3	5	4	7
9	1	5	4	2	6	3	7	8
3	2	6	8	7	1	9	5	4
8	7	4	5	3	9	2	6	1

EXPERT - 205

4	9	1	8	5	6	2	3	7
8	3	7	1	9	2	6	4	5
6	2	5	7	4	3	9	1	8
1	6	4	3	7	5	8	2	9
2	7	3	6	8	9	4	5	1
5	8	9	4	2	1	7	6	3
7	5	2	9	1	4	3	8	6
9	4	6	5	3	8	1	7	2
3	1	8	2	6	7	5	9	4

EXPERT - 206

5	7	9	2	3	1	6	4	8
1	8	6	5	4	9	3	2	7
3	2	4	6	7	8	9	1	5
9	3	2	1	5	7	4	8	6
4	1	7	8	2	6	5	9	3
8	6	5	4	9	3	2	7	1
2	4	3	7	8	5	1	6	9
7	5	1	9	6	4	8	3	2
6	9	8	3	1	2	7	5	4

EXPERT - 207

5	7	8	9	3	6	1	4	2
2	9	4	8	7	1	3	6	5
3	1	6	2	5	4	8	7	9
6	2	1	7	9	8	5	3	4
8	3	7	4	1	5	2	9	6
4	5	9	3	6	2	7	8	1
1	8	3	6	2	9	4	5	7
7	6	2	5	4	3	9	1	8
9	4	5	1	8	7	6	2	3

EXPERT - 208

1	3	8	9	7	5	4	6	2
7	9	2	6	4	8	5	1	3
5	4	6	3	1	2	8	9	7
3	6	9	4	5	1	2	7	8
4	2	1	7	8	6	3	5	9
8	5	7	2	9	3	1	4	6
9	1	4	8	2	7	6	3	5
2	7	3	5	6	4	9	8	1
6	8	5	1	3	9	7	2	4

EXPERT - 209

7	6	2	8	9	4	3	5	1
4	1	8	3	5	2	9	6	7
9	3	5	6	1	7	2	8	4
1	8	6	9	2	3	7	4	5
5	9	3	7	4	8	1	2	6
2	7	4	5	6	1	8	3	9
8	2	9	4	7	6	5	1	3
6	5	1	2	3	9	4	7	8
3	4	7	1	8	5	6	9	2

EXPERT - 210

2	5	6	9	4	7	1	8	3
3	1	8	2	6	5	7	4	9
9	4	7	8	3	1	5	2	6
6	8	1	7	5	3	2	9	4
4	7	2	6	8	9	3	1	5
5	9	3	1	2	4	8	6	7
7	2	5	4	9	8	6	3	1
1	6	9	3	7	2	4	5	8
8	3	4	5	1	6	9	7	2

EXPERT - 211

9	7	6	2	4	1	8	3	5
1	5	8	9	6	3	7	2	4
3	4	2	7	8	5	9	1	6
2	3	5	6	1	9	4	8	7
7	6	1	4	2	8	3	5	9
8	9	4	3	5	7	2	6	1
4	2	3	5	7	6	1	9	8
6	8	7	1	9	2	5	4	3
5	1	9	8	3	4	6	7	2

EXPERT - 212

8	3	6	4	5	7	2	1	9
5	7	4	1	9	2	6	8	3
9	2	1	3	6	8	5	7	4
7	6	9	8	3	4	1	2	5
2	1	8	9	7	5	3	4	6
4	5	3	2	1	6	8	9	7
1	4	5	7	8	3	9	6	2
6	9	7	5	2	1	4	3	8
3	8	2	6	4	9	7	5	1

EXPERT - 213

6	5	4	7	1	3	2	8	9
1	2	3	9	8	5	4	7	6
8	9	7	4	6	2	1	3	5
5	4	8	6	3	7	9	2	1
9	6	1	8	2	4	3	5	7
7	3	2	5	9	1	6	4	8
4	8	9	3	7	6	5	1	2
2	7	5	1	4	9	8	6	3
3	1	6	2	5	8	7	9	4

EXPERT - 214

3	8	7	5	4	1	9	2	6
6	5	9	2	3	8	1	4	7
4	1	2	9	7	6	3	8	5
7	9	3	4	8	2	5	6	1
8	2	5	1	6	9	7	3	4
1	6	4	3	5	7	8	9	2
2	3	1	7	9	4	6	5	8
9	4	6	8	1	5	2	7	3
5	7	8	6	2	3	4	1	9

EXPERT - 215

3	9	5	7	2	4	1	6	8
7	1	2	8	6	5	3	9	4
6	4	8	3	1	9	2	7	5
4	7	1	9	5	8	6	2	3
8	5	9	6	3	2	4	1	7
2	3	6	1	4	7	5	8	9
5	6	7	4	8	1	9	3	2
1	8	4	2	9	3	7	5	6
9	2	3	5	7	6	8	4	1

EXPERT - 216

7	4	2	5	6	1	9	8	3
8	5	1	9	2	3	6	4	7
9	3	6	4	8	7	5	2	1
4	2	3	8	7	5	1	6	9
6	1	9	2	3	4	8	7	5
5	8	7	6	1	9	4	3	2
2	7	4	1	5	8	3	9	6
1	6	8	3	9	2	7	5	4
3	9	5	7	4	6	2	1	8

EXPERT - 217

7	6	4	1	8	2	5	9	3
3	9	8	5	7	6	2	1	4
5	2	1	9	3	4	8	6	7
2	4	7	6	5	3	9	8	1
6	8	3	7	9	1	4	2	5
9	1	5	2	4	8	3	7	6
4	7	2	8	1	5	6	3	9
1	3	6	4	2	9	7	5	8
8	5	9	3	6	7	1	4	2

EXPERT - 218

7	2	6	3	5	8	9	4	1
9	4	1	2	7	6	8	3	5
8	5	3	4	9	1	6	7	2
6	8	7	1	3	4	5	2	9
4	3	2	9	8	5	7	1	6
1	9	5	6	2	7	4	8	3
3	7	8	5	1	9	2	6	4
5	1	4	8	6	2	3	9	7
2	6	9	7	4	3	1	5	8

EXPERT - 219

6	1	2	3	7	4	9	5	8
4	3	5	9	6	8	1	7	2
8	7	9	2	5	1	6	3	4
5	9	4	6	2	7	8	1	3
3	2	8	1	9	5	7	4	6
7	6	1	8	4	3	5	2	9
9	5	3	4	1	6	2	8	7
2	8	7	5	3	9	4	6	1
1	4	6	7	8	2	3	9	5

EXPERT - 220

9	8	3	4	6	7	1	5	2
4	1	7	8	5	2	3	6	9
6	5	2	1	3	9	7	8	4
7	4	8	9	1	6	2	3	5
2	3	9	5	4	8	6	7	1
5	6	1	2	7	3	4	9	8
8	7	4	6	2	5	9	1	3
1	9	6	3	8	4	5	2	7
3	2	5	7	9	1	8	4	6

EXPERT - 221

8	3	2	9	6	4	5	1	7
6	1	4	8	7	5	9	2	3
7	5	9	2	3	1	6	8	4
3	6	1	5	8	7	4	9	2
4	9	5	1	2	3	8	7	6
2	7	8	6	4	9	3	5	1
9	2	3	7	5	6	1	4	8
1	8	6	4	9	2	7	3	5
5	4	7	3	1	8	2	6	9

EXPERT - 222

3	6	4	2	8	9	1	7	5
9	1	7	3	4	5	6	8	2
8	5	2	1	7	6	9	4	3
1	9	6	5	3	4	8	2	7
4	2	5	7	9	8	3	6	1
7	8	3	6	2	1	5	9	4
2	3	9	8	1	7	4	5	6
6	4	1	9	5	2	7	3	8
5	7	8	4	6	3	2	1	9

EXPERT - 223

7	1	9	5	4	2	6	3	8
3	5	4	8	6	9	1	2	7
2	8	6	1	7	3	5	9	4
9	4	3	2	1	7	8	5	6
1	6	2	9	5	8	7	4	3
8	7	5	6	3	4	9	1	2
6	3	1	7	2	5	4	8	9
5	2	8	4	9	6	3	7	1
4	9	7	3	8	1	2	6	5

EXPERT - 224

4	8	1	6	7	5	3	9	2
7	9	2	1	3	4	8	5	6
6	3	5	8	2	9	7	1	4
2	7	4	5	8	3	9	6	1
3	1	9	4	6	7	5	2	8
8	5	6	2	9	1	4	7	3
5	6	8	7	4	2	1	3	9
1	4	3	9	5	6	2	8	7
9	2	7	3	1	8	6	4	5

EXPERT - 225

7	4	8	5	2	3	1	9	6
9	5	3	4	6	1	8	2	7
1	6	2	9	8	7	5	3	4
2	1	6	3	5	8	7	4	9
8	7	4	6	1	9	2	5	3
3	9	5	7	4	2	6	8	1
4	3	1	8	7	5	9	6	2
6	8	7	2	9	4	3	1	5
5	2	9	1	3	6	4	7	8

EXPERT - 226

3	6	7	4	9	1	5	2	8
4	1	5	2	7	8	3	6	9
9	8	2	3	6	5	4	1	7
8	2	3	9	1	7	6	5	4
1	9	6	5	4	2	7	8	3
7	5	4	6	8	3	1	9	2
6	3	1	7	2	9	8	4	5
2	7	8	1	5	4	9	3	6
5	4	9	8	3	6	2	7	1

EXPERT - 227

9	6	1	2	7	4	8	3	5
3	5	2	1	9	8	4	6	7
4	8	7	6	5	3	9	1	2
5	9	3	8	2	7	1	4	6
2	4	8	9	1	6	7	5	3
1	7	6	3	4	5	2	9	8
7	2	4	5	6	1	3	8	9
6	3	9	4	8	2	5	7	1
8	1	5	7	3	9	6	2	4

EXPERT - 228

4	5	9	8	6	1	7	2	3
2	7	1	4	3	5	8	6	9
3	8	6	2	9	7	1	5	4
1	4	8	6	5	9	3	7	2
6	9	2	7	8	3	5	4	1
7	3	5	1	4	2	6	9	8
8	6	7	9	1	4	2	3	5
9	2	3	5	7	8	4	1	6
5	1	4	3	2	6	9	8	7

EXPERT - 229

5	7	2	3	9	4	6	1	8
3	4	6	1	8	5	9	7	2
8	9	1	7	2	6	3	5	4
4	3	5	2	1	9	8	6	7
7	6	8	4	5	3	1	2	9
2	1	9	8	6	7	4	3	5
1	8	4	6	7	2	5	9	3
9	2	3	5	4	1	7	8	6
6	5	7	9	3	8	2	4	1

EXPERT - 230

4	7	5	3	9	2	8	1	6
1	8	3	6	7	5	2	9	4
6	2	9	4	1	8	3	5	7
5	3	2	9	8	4	6	7	1
9	1	7	2	6	3	4	8	5
8	6	4	7	5	1	9	3	2
2	9	1	8	4	7	5	6	3
7	4	6	5	3	9	1	2	8
3	5	8	1	2	6	7	4	9

EXPERT - 231

8	6	2	9	7	3	5	1	4
1	7	3	5	8	4	2	9	6
5	9	4	6	1	2	8	7	3
6	5	1	7	3	9	4	2	8
4	8	9	1	2	6	3	5	7
3	2	7	4	5	8	9	6	1
2	4	5	8	6	7	1	3	9
7	3	8	2	9	1	6	4	5
9	1	6	3	4	5	7	8	2

EXPERT - 232

3	7	6	2	5	4	9	8	1
8	2	5	6	1	9	7	4	3
9	4	1	3	7	8	5	6	2
2	6	7	1	8	3	4	5	9
4	1	8	5	9	7	3	2	6
5	3	9	4	6	2	1	7	8
6	8	3	7	4	1	2	9	5
7	5	2	9	3	6	8	1	4
1	9	4	8	2	5	6	3	7

EXPERT - 233

3	4	1	6	8	9	7	5	2
9	5	2	4	3	7	8	1	6
6	8	7	5	1	2	4	3	9
2	1	6	9	5	4	3	7	8
8	3	9	1	7	6	2	4	5
5	7	4	3	2	8	6	9	1
1	2	5	8	4	3	9	6	7
7	6	3	2	9	1	5	8	4
4	9	8	7	6	5	1	2	3

EXPERT - 234

6	9	8	1	7	4	5	2	3
4	1	3	5	2	9	7	8	6
5	2	7	3	8	6	1	9	4
9	6	1	4	3	2	8	5	7
2	8	4	9	5	7	3	6	1
3	7	5	6	1	8	9	4	2
7	5	9	2	6	1	4	3	8
8	4	2	7	9	3	6	1	5
1	3	6	8	4	5	2	7	9

EXPERT - 235

5	6	3	8	1	7	2	9	4
2	7	4	3	6	9	8	5	1
9	1	8	5	2	4	6	7	3
8	2	1	7	4	5	3	6	9
3	4	6	1	9	8	7	2	5
7	5	9	6	3	2	4	1	8
1	3	7	9	8	6	5	4	2
4	9	5	2	7	3	1	8	6
6	8	2	4	5	1	9	3	7

EXPERT - 236

8	3	2	5	7	4	1	6	9
6	7	9	8	3	1	4	2	5
1	4	5	6	9	2	3	8	7
2	5	4	9	1	8	6	7	3
7	8	1	3	2	6	5	9	4
9	6	3	7	4	5	8	1	2
4	2	8	1	5	7	9	3	6
3	1	7	4	6	9	2	5	8
5	9	6	2	8	3	7	4	1

EXPERT - 237

7	1	4	8	5	3	2	9	6
2	3	5	7	9	6	8	1	4
9	6	8	2	4	1	3	5	7
5	2	3	6	7	9	1	4	8
1	4	7	5	2	8	6	3	9
8	9	6	3	1	4	7	2	5
3	8	1	4	6	5	9	7	2
6	5	2	9	3	7	4	8	1
4	7	9	1	8	2	5	6	3

EXPERT - 238

7	3	2	9	6	5	4	1	8
4	5	8	1	3	2	7	9	6
9	1	6	7	4	8	5	2	3
5	4	1	8	9	6	3	7	2
3	6	7	2	5	4	9	8	1
2	8	9	3	7	1	6	4	5
1	7	3	6	8	9	2	5	4
6	2	5	4	1	7	8	3	9
8	9	4	5	2	3	1	6	7

EXPERT - 239

8	6	7	5	2	1	9	3	4
9	4	5	3	7	8	2	1	6
2	1	3	4	9	6	5	8	7
4	9	6	8	3	5	7	2	1
7	3	2	1	4	9	8	6	5
1	5	8	2	6	7	3	4	9
3	8	9	7	1	4	6	5	2
5	7	1	6	8	2	4	9	3
6	2	4	9	5	3	1	7	8

EXPERT - 240

7	5	3	6	8	1	9	4	2
9	4	8	5	2	3	7	6	1
6	1	2	9	4	7	3	5	8
4	3	6	2	1	8	5	7	9
5	9	1	3	7	6	8	2	4
8	2	7	4	9	5	6	1	3
3	8	5	1	6	2	4	9	7
1	7	4	8	5	9	2	3	6
2	6	9	7	3	4	1	8	5

EXPERT - 241

9	4	1	7	5	3	2	8	6
6	5	8	2	4	9	7	3	1
7	2	3	1	6	8	5	4	9
1	3	7	9	2	6	4	5	8
8	6	2	4	3	5	9	1	7
4	9	5	8	7	1	6	2	3
2	8	4	6	1	7	3	9	5
3	7	9	5	8	2	1	6	4
5	1	6	3	9	4	8	7	2

EXPERT - 242

8	1	7	2	6	5	9	3	4
2	4	3	1	9	7	5	6	8
6	9	5	4	8	3	7	2	1
3	5	2	8	7	6	1	4	9
4	7	8	9	2	1	3	5	6
1	6	9	5	3	4	2	8	7
9	8	1	3	4	2	6	7	5
5	2	6	7	1	8	4	9	3
7	3	4	6	5	9	8	1	2

EXPERT - 243

2	4	7	6	9	8	3	1	5
6	3	5	2	1	7	4	8	9
9	1	8	3	5	4	2	6	7
3	9	4	1	7	6	8	5	2
8	6	1	5	3	2	9	7	4
5	7	2	4	8	9	6	3	1
4	8	3	7	2	5	1	9	6
7	2	9	8	6	1	5	4	3
1	5	6	9	4	3	7	2	8

EXPERT - 244

3	7	1	5	2	6	8	9	4
9	4	8	7	3	1	6	5	2
6	2	5	8	4	9	7	1	3
4	1	7	6	8	2	9	3	5
2	5	6	3	9	7	1	4	8
8	9	3	4	1	5	2	7	6
7	6	2	1	5	4	3	8	9
1	8	4	9	6	3	5	2	7
5	3	9	2	7	8	4	6	1

EXPERT - 245

4	6	8	3	9	7	1	5	2
5	3	2	1	4	8	6	7	9
1	7	9	6	5	2	8	3	4
7	1	6	9	8	5	2	4	3
9	8	3	4	2	6	7	1	5
2	5	4	7	3	1	9	8	6
8	9	1	5	6	4	3	2	7
6	4	7	2	1	3	5	9	8
3	2	5	8	7	9	4	6	1

EXPERT - 246

6	7	9	5	4	8	3	1	2
3	4	1	9	2	6	5	7	8
5	8	2	1	7	3	4	9	6
1	9	4	6	8	5	2	3	7
2	6	5	3	9	7	8	4	1
8	3	7	4	1	2	9	6	5
7	2	3	8	6	4	1	5	9
4	1	6	2	5	9	7	8	3
9	5	8	7	3	1	6	2	4

EXPERT - 247

9	8	1	6	3	7	2	5	4
5	6	3	1	4	2	8	9	7
7	2	4	5	9	8	1	6	3
1	5	8	2	6	4	7	3	9
4	7	2	3	8	9	6	1	5
3	9	6	7	1	5	4	2	8
6	1	5	4	7	3	9	8	2
2	4	9	8	5	6	3	7	1
8	3	7	9	2	1	5	4	6

EXPERT - 248

4	1	8	3	2	5	9	7	6
3	7	9	6	4	8	2	1	5
5	2	6	9	7	1	4	3	8
2	4	1	5	6	3	7	8	9
7	9	5	1	8	4	6	2	3
6	8	3	2	9	7	5	4	1
8	5	2	4	1	6	3	9	7
1	6	4	7	3	9	8	5	2
9	3	7	8	5	2	1	6	4

EXPERT - 249

5	7	6	1	2	9	3	4	8
3	9	2	4	8	5	1	7	6
8	4	1	7	6	3	5	9	2
4	5	7	8	3	6	9	2	1
6	3	9	2	5	1	7	8	4
2	1	8	9	7	4	6	3	5
1	8	4	6	9	7	2	5	3
9	2	5	3	1	8	4	6	7
7	6	3	5	4	2	8	1	9

EXPERT - 250

7	6	2	3	5	1	8	4	9
3	5	4	8	6	9	1	7	2
8	9	1	4	7	2	5	3	6
9	4	8	7	2	5	6	1	3
5	1	7	9	3	6	2	8	4
6	2	3	1	4	8	9	5	7
4	8	5	6	9	7	3	2	1
2	3	6	5	1	4	7	9	8
1	7	9	2	8	3	4	6	5

EXPERT - 251

5	1	6	2	7	3	8	4	9
8	2	9	6	1	4	5	7	3
4	7	3	9	8	5	6	2	1
6	3	8	4	2	9	1	5	7
2	9	7	1	5	6	3	8	4
1	5	4	8	3	7	2	9	6
3	6	5	7	9	2	4	1	8
9	4	1	5	6	8	7	3	2
7	8	2	3	4	1	9	6	5

EXPERT - 252

5	4	1	7	3	6	2	9	8
3	2	7	9	1	8	4	5	6
8	6	9	5	2	4	1	3	7
2	5	6	8	4	9	7	1	3
1	7	4	2	5	3	6	8	9
9	3	8	1	6	7	5	4	2
6	1	3	4	9	2	8	7	5
4	8	2	3	7	5	9	6	1
7	9	5	6	8	1	3	2	4

EXPERT - 253

4	3	2	5	8	1	6	7	9
6	5	1	2	9	7	8	4	3
7	8	9	3	6	4	5	1	2
3	6	4	9	2	8	7	5	1
2	7	8	1	5	6	3	9	4
9	1	5	4	7	3	2	6	8
5	4	6	8	3	9	1	2	7
8	9	7	6	1	2	4	3	5
1	2	3	7	4	5	9	8	6

EXPERT - 254

1	9	3	7	2	8	6	5	4
6	8	2	1	5	4	3	9	7
5	4	7	6	9	3	8	2	1
9	1	4	5	7	6	2	3	8
3	2	5	8	4	9	7	1	6
8	7	6	2	3	1	9	4	5
7	5	8	3	1	2	4	6	9
4	3	1	9	6	7	5	8	2
2	6	9	4	8	5	1	7	3

EXPERT - 255

5	2	6	9	8	3	1	4	7
1	9	8	7	2	4	3	5	6
3	4	7	5	6	1	9	2	8
4	6	5	8	3	9	7	1	2
9	1	2	4	7	5	6	8	3
8	7	3	6	1	2	4	9	5
7	3	4	1	5	8	2	6	9
2	5	1	3	9	6	8	7	4
6	8	9	2	4	7	5	3	1

EXPERT - 256

9	3	8	7	2	1	4	6	5
7	1	6	4	3	5	8	2	9
5	4	2	9	8	6	1	3	7
2	8	4	5	6	7	3	9	1
3	5	7	2	1	9	6	8	4
6	9	1	8	4	3	5	7	2
8	7	9	3	5	4	2	1	6
1	2	5	6	9	8	7	4	3
4	6	3	1	7	2	9	5	8

EXPERT - 257

8	7	2	6	9	1	3	5	4
1	5	3	8	2	4	7	6	9
9	4	6	7	3	5	1	8	2
2	8	7	9	6	3	5	4	1
6	1	4	2	5	7	8	9	3
3	9	5	1	4	8	6	2	7
4	3	9	5	7	6	2	1	8
5	2	1	3	8	9	4	7	6
7	6	8	4	1	2	9	3	5

EXPERT - 258

6	3	4	1	9	7	2	5	8
5	2	1	4	3	8	9	6	7
9	7	8	5	6	2	4	3	1
4	9	6	7	5	3	1	8	2
8	5	7	2	1	4	3	9	6
3	1	2	6	8	9	5	7	4
1	8	3	9	4	6	7	2	5
2	6	5	3	7	1	8	4	9
7	4	9	8	2	5	6	1	3

EXPERT - 259

8	4	7	2	9	1	3	5	6
5	1	2	3	6	7	4	9	8
3	6	9	8	5	4	1	2	7
1	3	5	7	4	6	9	8	2
9	8	6	1	2	5	7	3	4
2	7	4	9	3	8	5	6	1
7	2	1	5	8	9	6	4	3
6	5	8	4	7	3	2	1	9
4	9	3	6	1	2	8	7	5

EXPERT - 260

1	3	2	6	4	9	5	7	8
7	6	8	1	5	3	9	2	4
4	9	5	7	2	8	3	6	1
3	4	7	8	6	2	1	5	9
8	2	9	5	3	1	6	4	7
6	5	1	4	9	7	8	3	2
5	8	6	2	1	4	7	9	3
9	1	4	3	7	6	2	8	5
2	7	3	9	8	5	4	1	6

EXPERT - 261

1	5	2	4	8	3	6	7	9
9	8	3	1	7	6	4	2	5
4	7	6	5	9	2	8	1	3
5	4	1	7	2	9	3	6	8
6	3	7	8	4	1	9	5	2
8	2	9	3	6	5	1	4	7
7	1	8	9	5	4	2	3	6
3	6	5	2	1	8	7	9	4
2	9	4	6	3	7	5	8	1

EXPERT - 262

3	5	7	4	8	2	6	9	1
4	6	8	7	1	9	5	3	2
1	2	9	3	5	6	8	7	4
7	1	6	2	4	3	9	8	5
8	3	2	9	7	5	1	4	6
5	9	4	8	6	1	3	2	7
6	8	3	5	2	7	4	1	9
2	4	5	1	9	8	7	6	3
9	7	1	6	3	4	2	5	8

EXPERT - 263

5	8	4	6	1	7	9	3	2
2	9	3	8	5	4	1	6	7
1	7	6	9	3	2	5	4	8
8	5	2	7	6	1	3	9	4
9	3	1	2	4	5	7	8	6
6	4	7	3	8	9	2	5	1
3	2	8	1	9	6	4	7	5
4	1	9	5	7	8	6	2	3
7	6	5	4	2	3	8	1	9

EXPERT - 264

6	1	7	3	9	8	2	5	4
9	5	3	4	2	7	6	8	1
4	8	2	5	1	6	3	7	9
1	7	9	6	8	4	5	3	2
3	2	4	1	7	5	8	9	6
5	6	8	9	3	2	1	4	7
7	3	6	8	4	1	9	2	5
8	4	5	2	6	9	7	1	3
2	9	1	7	5	3	4	6	8

EXPERT - 265

3	7	6	4	5	1	9	8	2
8	5	2	9	3	6	4	7	1
9	1	4	8	2	7	5	3	6
6	2	1	7	9	4	3	5	8
5	3	9	1	8	2	6	4	7
4	8	7	3	6	5	2	1	9
2	4	8	5	1	9	7	6	3
1	9	5	6	7	3	8	2	4
7	6	3	2	4	8	1	9	5

EXPERT - 266

8	9	7	4	5	6	3	1	2
5	4	6	1	2	3	8	7	9
1	2	3	8	9	7	6	4	5
2	8	9	7	3	1	5	6	4
7	3	5	6	4	2	9	8	1
4	6	1	9	8	5	2	3	7
9	5	8	3	1	4	7	2	6
3	7	4	2	6	9	1	5	8
6	1	2	5	7	8	4	9	3

EXPERT - 267

2	7	3	6	4	5	9	1	8
9	4	8	1	2	3	7	5	6
5	1	6	8	7	9	4	2	3
1	3	9	4	5	6	2	8	7
4	2	5	7	9	8	3	6	1
6	8	7	2	3	1	5	9	4
7	6	4	5	1	2	8	3	9
3	5	1	9	8	7	6	4	2
8	9	2	3	6	4	1	7	5

EXPERT - 268

8	9	1	2	3	7	6	4	5
4	3	5	9	8	6	1	2	7
6	2	7	1	5	4	3	8	9
9	8	6	5	1	2	4	7	3
3	7	2	4	6	9	8	5	1
1	5	4	3	7	8	2	9	6
7	1	8	6	2	5	9	3	4
2	6	9	7	4	3	5	1	8
5	4	3	8	9	1	7	6	2

EXPERT - 269

6	7	3	1	9	5	2	4	8
2	8	9	4	6	7	3	1	5
5	1	4	2	8	3	7	9	6
7	5	1	6	2	4	9	8	3
3	4	8	9	5	1	6	7	2
9	2	6	7	3	8	4	5	1
1	3	2	8	7	9	5	6	4
8	9	5	3	4	6	1	2	7
4	6	7	5	1	2	8	3	9

EXPERT - 270

1	5	6	2	4	9	3	7	8
4	9	3	5	8	7	6	1	2
2	8	7	1	6	3	5	4	9
5	7	9	8	1	6	4	2	3
6	2	4	7	3	5	9	8	1
3	1	8	4	9	2	7	5	6
8	6	1	9	7	4	2	3	5
7	3	2	6	5	8	1	9	4
9	4	5	3	2	1	8	6	7

EXPERT - 271

7	5	8	3	9	4	2	6	1
2	9	6	1	8	7	4	3	5
3	1	4	5	2	6	9	8	7
4	8	2	6	7	5	3	1	9
1	6	7	9	3	2	8	5	4
5	3	9	4	1	8	6	7	2
9	2	3	7	6	1	5	4	8
6	4	1	8	5	9	7	2	3
8	7	5	2	4	3	1	9	6

EXPERT - 272

8	2	3	1	4	6	5	7	9
1	7	5	9	2	8	6	3	4
4	6	9	7	3	5	2	1	8
5	9	1	8	6	2	3	4	7
7	8	4	5	9	3	1	2	6
2	3	6	4	1	7	8	9	5
9	5	2	3	8	4	7	6	1
6	1	7	2	5	9	4	8	3
3	4	8	6	7	1	9	5	2

EXPERT - 273

8	9	1	5	3	2	6	7	4
7	3	4	9	6	1	8	5	2
2	5	6	7	4	8	1	9	3
1	2	5	3	9	4	7	8	6
6	8	9	1	2	7	3	4	5
3	4	7	8	5	6	2	1	9
5	6	8	4	7	3	9	2	1
4	1	2	6	8	9	5	3	7
9	7	3	2	1	5	4	6	8

EXPERT - 274

9	1	8	5	4	6	7	3	2
6	2	4	1	7	3	9	5	8
7	3	5	8	9	2	6	4	1
8	5	7	6	1	4	2	9	3
3	9	6	2	5	7	1	8	4
2	4	1	3	8	9	5	6	7
5	8	3	9	2	1	4	7	6
1	7	9	4	6	8	3	2	5
4	6	2	7	3	5	8	1	9

EXPERT - 275

5	6	2	8	9	4	7	1	3
3	7	8	5	1	2	9	6	4
9	4	1	7	6	3	8	5	2
4	3	6	2	5	9	1	8	7
1	2	9	3	8	7	6	4	5
8	5	7	6	4	1	3	2	9
6	8	4	9	7	5	2	3	1
2	9	5	1	3	8	4	7	6
7	1	3	4	2	6	5	9	8

EXPERT - 276

5	6	4	2	7	9	3	1	8
7	1	9	8	4	3	5	6	2
3	2	8	5	1	6	7	4	9
6	3	7	9	8	1	4	2	5
1	8	2	4	5	7	6	9	3
9	4	5	6	3	2	1	8	7
4	9	3	1	2	5	8	7	6
2	5	1	7	6	8	9	3	4
8	7	6	3	9	4	2	5	1

EXPERT - 277

9	2	4	8	7	3	6	5	1
1	5	3	4	2	6	8	7	9
7	8	6	9	1	5	2	4	3
8	6	9	2	5	4	3	1	7
2	7	5	1	3	8	4	9	6
3	4	1	6	9	7	5	8	2
6	1	7	5	4	2	9	3	8
5	3	8	7	6	9	1	2	4
4	9	2	3	8	1	7	6	5

EXPERT - 278

1	5	2	6	4	8	3	9	7
7	6	9	1	3	5	2	8	4
3	4	8	7	9	2	1	5	6
5	1	4	2	8	7	9	6	3
9	3	6	4	5	1	7	2	8
2	8	7	3	6	9	5	4	1
6	7	1	9	2	4	8	3	5
4	9	5	8	7	3	6	1	2
8	2	3	5	1	6	4	7	9

EXPERT - 279

7	1	5	9	2	3	4	6	8
2	3	6	8	4	1	5	9	7
4	9	8	5	6	7	1	3	2
3	8	9	4	1	6	2	7	5
5	4	2	7	9	8	3	1	6
6	7	1	3	5	2	9	8	4
8	2	4	1	7	9	6	5	3
1	6	3	2	8	5	7	4	9
9	5	7	6	3	4	8	2	1

EXPERT - 280

2	8	4	7	5	6	9	3	1
3	5	6	1	4	9	2	8	7
1	9	7	8	3	2	6	4	5
7	2	9	4	6	1	3	5	8
4	1	5	3	9	8	7	6	2
8	6	3	2	7	5	4	1	9
6	4	2	5	8	7	1	9	3
9	7	8	6	1	3	5	2	4
5	3	1	9	2	4	8	7	6

EXPERT - 281

4	5	8	3	1	9	2	6	7
6	3	9	4	7	2	5	8	1
7	2	1	6	8	5	9	4	3
1	8	3	9	5	7	6	2	4
5	7	4	2	6	3	1	9	8
9	6	2	8	4	1	7	3	5
8	4	7	5	9	6	3	1	2
2	1	6	7	3	8	4	5	9
3	9	5	1	2	4	8	7	6

EXPERT - 282

8	1	3	2	9	6	5	7	4
5	2	6	1	7	4	8	3	9
9	7	4	3	8	5	2	1	6
2	8	1	5	6	3	4	9	7
3	4	5	9	2	7	6	8	1
6	9	7	8	4	1	3	5	2
4	5	8	7	1	2	9	6	3
1	3	2	6	5	9	7	4	8
7	6	9	4	3	8	1	2	5

EXPERT - 283

2	4	6	3	1	5	7	8	9
7	8	5	6	9	4	3	2	1
3	1	9	7	2	8	6	5	4
9	3	8	1	4	2	5	7	6
4	2	1	5	7	6	8	9	3
5	6	7	9	8	3	4	1	2
1	5	4	2	6	7	9	3	8
6	9	3	8	5	1	2	4	7
8	7	2	4	3	9	1	6	5

EXPERT - 284

5	8	3	4	9	1	2	6	7
9	6	4	2	3	7	5	8	1
7	1	2	8	6	5	9	4	3
1	5	8	9	7	2	4	3	6
4	9	7	3	8	6	1	5	2
3	2	6	1	5	4	8	7	9
6	4	1	7	2	8	3	9	5
2	3	5	6	4	9	7	1	8
8	7	9	5	1	3	6	2	4

EXPERT - 285

6	8	5	3	7	2	4	9	1
2	1	3	4	9	5	8	6	7
7	4	9	1	6	8	2	3	5
5	6	2	9	3	1	7	8	4
8	9	1	7	2	4	6	5	3
3	7	4	8	5	6	9	1	2
9	5	7	6	4	3	1	2	8
4	3	8	2	1	9	5	7	6
1	2	6	5	8	7	3	4	9

EXPERT - 286

2	7	1	4	3	6	5	9	8
6	9	3	1	5	8	2	7	4
4	5	8	7	2	9	3	6	1
3	8	9	5	1	2	7	4	6
5	1	4	8	6	7	9	2	3
7	2	6	9	4	3	1	8	5
8	4	7	3	9	5	6	1	2
9	3	2	6	8	1	4	5	7
1	6	5	2	7	4	8	3	9

EXPERT - 287

6	3	5	7	1	9	2	4	8
4	9	1	2	8	3	5	6	7
7	8	2	4	5	6	1	9	3
8	1	6	3	4	2	9	7	5
2	5	7	6	9	8	3	1	4
9	4	3	1	7	5	8	2	6
5	7	8	9	2	4	6	3	1
3	2	4	8	6	1	7	5	9
1	6	9	5	3	7	4	8	2

EXPERT - 288

5	3	1	2	4	8	9	6	7
7	6	8	1	9	5	2	3	4
2	9	4	6	7	3	8	5	1
6	2	5	7	3	4	1	9	8
8	7	3	9	5	1	4	2	6
4	1	9	8	6	2	5	7	3
1	5	2	3	8	7	6	4	9
9	4	7	5	1	6	3	8	2
3	8	6	4	2	9	7	1	5

EXPERT - 289

5	4	3	9	8	7	6	1	2
6	8	9	1	2	3	7	5	4
2	1	7	6	4	5	8	9	3
9	2	1	4	3	8	5	7	6
8	5	4	7	6	2	9	3	1
7	3	6	5	9	1	4	2	8
4	7	5	2	1	6	3	8	9
3	9	2	8	7	4	1	6	5
1	6	8	3	5	9	2	4	7

EXPERT - 290

5	9	3	8	2	6	1	7	4
7	4	6	5	9	1	3	8	2
1	2	8	4	7	3	6	5	9
2	7	4	6	8	9	5	1	3
8	3	1	7	5	4	2	9	6
9	6	5	1	3	2	7	4	8
3	1	2	9	4	7	8	6	5
4	8	7	3	6	5	9	2	1
6	5	9	2	1	8	4	3	7

EXPERT - 291

2	6	8	4	3	7	9	5	1
7	5	4	9	2	1	8	6	3
3	1	9	5	8	6	7	2	4
9	8	1	2	6	3	4	7	5
5	7	6	8	9	4	1	3	2
4	3	2	1	7	5	6	9	8
1	9	7	3	4	2	5	8	6
6	4	3	7	5	8	2	1	9
8	2	5	6	1	9	3	4	7

EXPERT - 292

8	7	9	4	3	6	1	2	5
3	4	2	1	5	8	6	7	9
6	1	5	9	2	7	8	4	3
9	5	8	6	7	1	4	3	2
4	3	6	2	9	5	7	8	1
1	2	7	8	4	3	5	9	6
7	9	4	5	6	2	3	1	8
2	6	1	3	8	4	9	5	7
5	8	3	7	1	9	2	6	4

EXPERT - 293

3	8	5	9	7	4	6	2	1
2	7	1	5	3	6	4	9	8
6	9	4	8	1	2	5	3	7
7	2	9	6	4	8	3	1	5
4	1	8	3	2	5	7	6	9
5	3	6	1	9	7	2	8	4
1	6	7	2	5	9	8	4	3
9	5	2	4	8	3	1	7	6
8	4	3	7	6	1	9	5	2

EXPERT - 294

9	5	4	6	7	1	2	3	8
1	2	6	9	3	8	4	5	7
3	7	8	5	2	4	9	6	1
5	8	1	3	4	2	7	9	6
4	9	7	1	5	6	3	8	2
6	3	2	8	9	7	5	1	4
2	6	9	7	1	3	8	4	5
8	4	5	2	6	9	1	7	3
7	1	3	4	8	5	6	2	9

EXPERT - 295

8	9	7	3	2	4	1	6	5
6	5	3	8	9	1	7	4	2
2	4	1	7	5	6	9	8	3
9	1	2	5	6	8	4	3	7
5	3	4	9	7	2	6	1	8
7	8	6	1	4	3	2	5	9
1	2	5	4	8	7	3	9	6
3	6	9	2	1	5	8	7	4
4	7	8	6	3	9	5	2	1

EXPERT - 296

1	9	8	5	6	7	4	2	3
6	2	7	4	1	3	8	9	5
4	5	3	2	8	9	7	6	1
9	6	5	7	4	8	1	3	2
7	1	4	3	2	6	5	8	9
8	3	2	1	9	5	6	4	7
2	8	6	9	5	1	3	7	4
5	7	9	8	3	4	2	1	6
3	4	1	6	7	2	9	5	8

EXPERT - 297

9	8	1	3	5	2	6	4	7
2	4	3	8	7	6	5	9	1
6	5	7	4	1	9	2	3	8
8	1	2	5	6	4	9	7	3
4	7	9	1	2	3	8	6	5
3	6	5	7	9	8	1	2	4
1	9	8	6	3	7	4	5	2
7	2	4	9	8	5	3	1	6
5	3	6	2	4	1	7	8	9

EXPERT - 298

8	2	1	4	3	9	6	7	5
4	5	9	1	7	6	2	8	3
7	3	6	5	8	2	4	1	9
2	7	8	9	4	3	5	6	1
6	1	4	7	5	8	3	9	2
3	9	5	2	6	1	7	4	8
5	8	3	6	9	4	1	2	7
9	6	2	3	1	7	8	5	4
1	4	7	8	2	5	9	3	6

EXPERT - 299

9	3	2	7	6	8	1	4	5
7	8	4	3	5	1	6	9	2
5	6	1	4	9	2	3	7	8
8	4	6	2	3	5	7	1	9
1	2	7	9	4	6	5	8	3
3	5	9	1	8	7	2	6	4
2	1	5	8	7	9	4	3	6
4	7	8	6	2	3	9	5	1
6	9	3	5	1	4	8	2	7

EXPERT - 300

6	5	8	3	1	4	9	7	2
7	4	9	6	5	2	3	8	1
3	2	1	8	7	9	5	4	6
4	7	2	1	6	5	8	3	9
8	9	6	2	4	3	7	1	5
5	1	3	7	9	8	2	6	4
1	8	7	9	2	6	4	5	3
9	3	4	5	8	1	6	2	7
2	6	5	4	3	7	1	9	8

EXPERT - 301

7	3	5	6	1	9	4	8	2
9	2	6	8	7	4	1	5	3
1	8	4	3	2	5	6	7	9
6	5	8	1	4	3	2	9	7
3	9	7	2	5	6	8	1	4
2	4	1	7	9	8	5	3	6
4	7	9	5	8	2	3	6	1
5	6	2	9	3	1	7	4	8
8	1	3	4	6	7	9	2	5

EXPERT - 302

3	9	5	1	8	6	2	4	7
7	1	8	2	5	4	6	9	3
6	4	2	7	3	9	8	1	5
1	8	4	3	2	7	9	5	6
2	6	9	8	4	5	7	3	1
5	3	7	6	9	1	4	8	2
8	7	6	9	1	3	5	2	4
4	2	1	5	7	8	3	6	9
9	5	3	4	6	2	1	7	8

EXPERT - 303

4	5	9	3	1	8	2	7	6
8	1	6	7	9	2	4	5	3
3	7	2	5	4	6	9	1	8
1	6	7	2	3	9	8	4	5
9	3	5	8	6	4	1	2	7
2	4	8	1	7	5	6	3	9
7	8	4	6	5	1	3	9	2
5	2	1	9	8	3	7	6	4
6	9	3	4	2	7	5	8	1

EXPERT - 304

5	7	9	8	6	4	1	2	3
4	1	8	3	2	7	6	9	5
2	6	3	1	9	5	4	8	7
3	4	7	6	8	2	9	5	1
1	5	6	7	3	9	2	4	8
8	9	2	5	4	1	7	3	6
9	8	5	2	7	6	3	1	4
7	2	1	4	5	3	8	6	9
6	3	4	9	1	8	5	7	2

EXPERT - 305

4	9	2	1	8	5	3	7	6
7	1	8	3	4	6	5	2	9
5	6	3	7	2	9	8	1	4
1	4	6	9	5	7	2	3	8
3	8	7	2	1	4	6	9	5
9	2	5	8	6	3	1	4	7
6	7	9	5	3	1	4	8	2
2	5	1	4	7	8	9	6	3
8	3	4	6	9	2	7	5	1

EXPERT - 306

9	5	7	1	8	3	2	6	4
3	2	4	5	6	9	7	8	1
6	8	1	7	4	2	5	3	9
8	6	5	2	9	7	4	1	3
4	9	2	8	3	1	6	5	7
1	7	3	4	5	6	9	2	8
2	1	6	9	7	8	3	4	5
7	4	8	3	2	5	1	9	6
5	3	9	6	1	4	8	7	2

EXPERT - 307

3	4	8	6	9	7	2	5	1
7	1	2	5	8	3	9	6	4
6	5	9	4	1	2	7	3	8
1	6	4	2	7	8	3	9	5
8	3	5	9	4	6	1	7	2
2	9	7	1	3	5	8	4	6
4	7	3	8	6	1	5	2	9
9	2	1	7	5	4	6	8	3
5	8	6	3	2	9	4	1	7

EXPERT - 308

1	6	5	9	8	4	2	3	7
7	2	4	5	1	3	9	6	8
8	3	9	2	7	6	5	4	1
4	5	2	6	9	7	1	8	3
9	1	6	8	3	5	7	2	4
3	8	7	1	4	2	6	9	5
5	9	3	4	2	1	8	7	6
6	7	8	3	5	9	4	1	2
2	4	1	7	6	8	3	5	9

EXPERT - 309

2	5	7	1	6	8	4	9	3
8	3	1	7	4	9	2	5	6
4	6	9	3	2	5	8	1	7
6	1	2	4	8	3	5	7	9
7	9	5	6	1	2	3	4	8
3	4	8	5	9	7	6	2	1
9	8	4	2	3	1	7	6	5
5	2	3	9	7	6	1	8	4
1	7	6	8	5	4	9	3	2

EXPERT - 310

1	9	8	4	6	2	5	3	7
4	7	6	8	3	5	2	9	1
2	5	3	7	1	9	4	6	8
8	4	2	9	7	3	6	1	5
9	3	5	6	4	1	8	7	2
7	6	1	2	5	8	3	4	9
6	2	9	1	8	4	7	5	3
3	1	7	5	2	6	9	8	4
5	8	4	3	9	7	1	2	6

EXPERT - 311

3	9	7	5	6	8	4	1	2
2	4	1	9	3	7	6	8	5
8	6	5	2	4	1	7	3	9
6	5	8	1	7	9	2	4	3
7	3	2	6	5	4	1	9	8
4	1	9	8	2	3	5	6	7
1	8	6	7	9	2	3	5	4
9	7	3	4	1	5	8	2	6
5	2	4	3	8	6	9	7	1

EXPERT - 312

1	4	8	9	6	2	5	7	3
3	9	6	5	7	8	1	4	2
7	5	2	1	4	3	9	6	8
8	3	4	6	5	9	2	1	7
9	2	1	3	8	7	6	5	4
5	6	7	2	1	4	3	8	9
6	8	9	7	2	1	4	3	5
4	1	3	8	9	5	7	2	6
2	7	5	4	3	6	8	9	1

EXPERT - 313

9	2	7	8	1	5	3	4	6
3	6	5	4	2	7	1	9	8
8	1	4	6	9	3	7	5	2
4	5	2	3	7	8	6	1	9
7	3	9	2	6	1	5	8	4
6	8	1	9	5	4	2	3	7
1	4	6	5	8	2	9	7	3
2	7	8	1	3	9	4	6	5
5	9	3	7	4	6	8	2	1

EXPERT - 314

9	2	7	4	8	3	6	1	5
3	5	1	7	6	9	8	4	2
8	4	6	1	2	5	7	3	9
7	9	4	5	3	8	2	6	1
5	1	8	6	7	2	4	9	3
6	3	2	9	4	1	5	7	8
1	7	9	8	5	4	3	2	6
4	8	3	2	1	6	9	5	7
2	6	5	3	9	7	1	8	4

EXPERT - 315

2	5	4	6	9	3	1	8	7
9	1	3	5	7	8	2	6	4
8	7	6	4	1	2	5	3	9
1	4	7	3	5	9	6	2	8
3	9	8	2	6	4	7	5	1
5	6	2	7	8	1	9	4	3
6	3	9	8	2	7	4	1	5
4	2	1	9	3	5	8	7	6
7	8	5	1	4	6	3	9	2

EXPERT - 316

5	2	9	7	1	3	6	4	8
7	3	4	6	8	9	1	5	2
6	8	1	2	4	5	7	9	3
2	5	6	8	7	1	4	3	9
9	4	8	5	3	6	2	7	1
1	7	3	9	2	4	5	8	6
8	9	5	1	6	7	3	2	4
4	1	7	3	9	2	8	6	5
3	6	2	4	5	8	9	1	7

EXPERT - 317

5	7	8	3	4	6	9	1	2
3	1	6	2	9	5	8	7	4
2	4	9	7	8	1	5	3	6
6	5	7	9	1	8	2	4	3
4	9	3	6	7	2	1	5	8
8	2	1	5	3	4	7	6	9
7	8	5	4	2	3	6	9	1
9	3	2	1	6	7	4	8	5
1	6	4	8	5	9	3	2	7

EXPERT - 318

4	5	1	8	7	9	6	2	3
2	8	6	4	3	1	5	7	9
7	3	9	2	5	6	8	1	4
5	9	4	7	6	2	1	3	8
1	2	7	3	4	8	9	6	5
8	6	3	9	1	5	2	4	7
9	4	8	1	2	3	7	5	6
6	7	2	5	9	4	3	8	1
3	1	5	6	8	7	4	9	2

EXPERT - 319

4	9	2	6	5	8	3	7	1
5	3	1	2	7	9	4	8	6
7	6	8	4	1	3	9	5	2
6	7	9	8	3	5	2	1	4
8	1	3	7	4	2	5	6	9
2	4	5	1	9	6	8	3	7
3	5	4	9	6	1	7	2	8
1	8	7	3	2	4	6	9	5
9	2	6	5	8	7	1	4	3

EXPERT - 320

8	6	2	7	9	4	5	3	1
4	5	1	3	8	6	9	2	7
9	7	3	1	5	2	4	6	8
1	4	7	5	2	8	6	9	3
2	3	5	6	7	9	8	1	4
6	9	8	4	1	3	2	7	5
7	1	4	2	6	5	3	8	9
5	2	9	8	3	7	1	4	6
3	8	6	9	4	1	7	5	2

EXPERT - 321

2	3	8	5	7	6	1	9	4
1	6	5	4	2	9	3	7	8
9	7	4	3	1	8	6	2	5
4	8	7	1	6	2	5	3	9
6	5	2	9	3	4	7	8	1
3	1	9	7	8	5	2	4	6
8	4	1	2	5	7	9	6	3
5	2	6	8	9	3	4	1	7
7	9	3	6	4	1	8	5	2

EXPERT - 322

4	8	3	1	9	6	2	7	5
2	9	6	7	5	4	1	8	3
1	5	7	8	3	2	4	6	9
8	7	1	2	6	9	5	3	4
3	2	5	4	8	1	7	9	6
9	6	4	3	7	5	8	2	1
7	4	8	6	1	3	9	5	2
5	3	2	9	4	7	6	1	8
6	1	9	5	2	8	3	4	7

EXPERT - 323

5	7	4	3	9	8	1	6	2
2	8	1	6	4	5	7	3	9
3	9	6	1	7	2	4	5	8
6	5	3	4	8	9	2	7	1
7	1	9	2	3	6	8	4	5
4	2	8	5	1	7	3	9	6
1	6	5	7	2	4	9	8	3
8	3	7	9	6	1	5	2	4
9	4	2	8	5	3	6	1	7

EXPERT - 324

9	3	6	2	4	7	5	8	1
7	1	4	3	8	5	2	6	9
8	2	5	1	6	9	7	3	4
5	8	9	7	3	4	1	2	6
2	4	1	6	9	8	3	7	5
6	7	3	5	1	2	4	9	8
4	5	2	8	7	6	9	1	3
3	9	8	4	2	1	6	5	7
1	6	7	9	5	3	8	4	2

EXPERT - 325

8	2	1	6	4	7	3	9	5
4	6	7	5	3	9	8	1	2
9	5	3	1	2	8	4	7	6
6	8	4	2	9	5	1	3	7
3	9	5	8	7	1	2	6	4
1	7	2	3	6	4	5	8	9
7	3	8	9	5	2	6	4	1
2	1	9	4	8	6	7	5	3
5	4	6	7	1	3	9	2	8

EXPERT - 326

6	7	4	9	5	2	1	8	3
1	9	2	8	6	3	5	4	7
5	8	3	1	4	7	2	6	9
4	3	8	2	1	6	7	9	5
9	5	7	3	8	4	6	2	1
2	6	1	7	9	5	4	3	8
7	2	5	6	3	8	9	1	4
8	1	6	4	7	9	3	5	2
3	4	9	5	2	1	8	7	6

EXPERT - 327

3	5	9	7	4	6	2	8	1
2	7	8	1	9	3	5	6	4
6	4	1	2	8	5	9	7	3
7	6	3	9	5	2	4	1	8
4	1	5	8	6	7	3	2	9
9	8	2	3	1	4	7	5	6
5	9	4	6	7	8	1	3	2
1	3	6	5	2	9	8	4	7
8	2	7	4	3	1	6	9	5

EXPERT - 328

4	1	9	6	8	3	7	5	2
7	8	5	1	4	2	6	9	3
6	2	3	9	7	5	4	1	8
5	4	2	7	3	1	9	8	6
1	3	6	8	2	9	5	7	4
8	9	7	4	5	6	2	3	1
2	6	1	3	9	7	8	4	5
9	5	8	2	1	4	3	6	7
3	7	4	5	6	8	1	2	9

EXPERT - 329

3	9	8	4	7	2	6	5	1
6	5	2	1	8	9	7	4	3
7	1	4	3	5	6	2	8	9
4	7	6	9	2	5	1	3	8
2	8	5	6	3	1	4	9	7
9	3	1	8	4	7	5	6	2
5	6	9	7	1	3	8	2	4
1	4	3	2	6	8	9	7	5
8	2	7	5	9	4	3	1	6

EXPERT - 330

8	9	2	7	3	6	1	4	5
7	6	1	4	5	2	9	3	8
4	5	3	8	9	1	7	6	2
3	7	6	1	2	5	8	9	4
1	4	5	9	7	8	6	2	3
2	8	9	6	4	3	5	1	7
6	1	4	2	8	7	3	5	9
5	2	8	3	6	9	4	7	1
9	3	7	5	1	4	2	8	6

EXPERT - 331

9	6	8	2	3	4	5	1	7
1	5	2	7	9	6	8	3	4
3	7	4	5	8	1	6	9	2
8	4	1	9	7	3	2	5	6
2	9	5	6	4	8	1	7	3
7	3	6	1	5	2	4	8	9
5	8	3	4	6	9	7	2	1
4	2	7	3	1	5	9	6	8
6	1	9	8	2	7	3	4	5

EXPERT - 332

9	5	6	4	3	2	1	7	8
2	4	8	1	7	5	9	6	3
3	7	1	8	9	6	4	2	5
8	6	9	7	2	3	5	1	4
7	3	4	5	1	9	2	8	6
1	2	5	6	4	8	7	3	9
5	9	2	3	6	1	8	4	7
6	1	7	9	8	4	3	5	2
4	8	3	2	5	7	6	9	1

EXPERT - 333

1	9	6	8	4	7	2	3	5
5	4	2	3	1	6	7	8	9
8	3	7	2	9	5	6	4	1
7	5	1	6	2	8	4	9	3
4	6	3	1	7	9	5	2	8
9	2	8	5	3	4	1	7	6
6	7	4	9	8	1	3	5	2
3	8	5	4	6	2	9	1	7
2	1	9	7	5	3	8	6	4

EXPERT - 334

4	5	1	7	8	2	6	3	9
8	2	6	5	3	9	7	1	4
3	7	9	4	6	1	2	8	5
5	9	2	3	4	6	8	7	1
1	3	4	8	2	7	5	9	6
6	8	7	9	1	5	3	4	2
7	6	8	1	5	4	9	2	3
2	4	3	6	9	8	1	5	7
9	1	5	2	7	3	4	6	8

EXPERT - 335

1	2	6	7	8	3	4	9	5
3	4	8	5	6	9	2	7	1
5	9	7	4	1	2	3	8	6
6	3	5	9	7	8	1	4	2
7	1	4	6	2	5	8	3	9
9	8	2	1	3	4	6	5	7
2	7	9	8	4	6	5	1	3
4	6	1	3	5	7	9	2	8
8	5	3	2	9	1	7	6	4

EXPERT - 336

5	8	3	9	6	7	2	1	4
6	7	9	2	1	4	8	5	3
4	2	1	8	3	5	6	9	7
2	1	6	3	5	8	4	7	9
9	5	7	6	4	1	3	8	2
8	3	4	7	2	9	1	6	5
1	6	2	5	7	3	9	4	8
3	9	5	4	8	6	7	2	1
7	4	8	1	9	2	5	3	6

EXPERT - 337

7	8	9	1	2	4	5	3	6
4	6	1	7	3	5	9	8	2
5	3	2	6	9	8	7	4	1
9	1	7	4	6	3	8	2	5
3	2	4	8	5	9	1	6	7
6	5	8	2	7	1	4	9	3
8	4	5	3	1	6	2	7	9
2	9	3	5	4	7	6	1	8
1	7	6	9	8	2	3	5	4

EXPERT - 338

1	6	2	7	3	9	8	4	5
3	8	5	4	1	2	6	7	9
4	7	9	6	8	5	3	1	2
7	9	6	2	5	3	4	8	1
2	3	1	8	4	7	5	9	6
8	5	4	1	9	6	7	2	3
9	4	7	3	6	1	2	5	8
6	1	8	5	2	4	9	3	7
5	2	3	9	7	8	1	6	4

EXPERT - 339

7	6	9	1	4	2	5	3	8
3	2	5	9	8	6	1	4	7
8	1	4	7	3	5	9	2	6
4	5	6	8	7	3	2	9	1
1	9	8	2	6	4	3	7	5
2	3	7	5	9	1	8	6	4
6	4	2	3	1	8	7	5	9
5	7	1	4	2	9	6	8	3
9	8	3	6	5	7	4	1	2

EXPERT - 340

5	7	6	1	4	8	2	9	3
9	4	3	7	6	2	8	1	5
8	2	1	3	9	5	4	7	6
4	3	8	5	7	9	6	2	1
7	5	2	4	1	6	9	3	8
1	6	9	8	2	3	5	4	7
2	9	5	6	3	7	1	8	4
3	8	4	9	5	1	7	6	2
6	1	7	2	8	4	3	5	9

EXPERT - 341

9	1	5	2	8	4	3	6	7
4	8	3	6	9	7	1	2	5
7	2	6	3	1	5	8	9	4
2	7	8	9	5	6	4	1	3
5	3	1	4	2	8	6	7	9
6	9	4	7	3	1	5	8	2
1	5	7	8	4	9	2	3	6
8	6	2	5	7	3	9	4	1
3	4	9	1	6	2	7	5	8

EXPERT - 342

2	7	6	4	3	5	8	9	1
5	3	8	6	9	1	4	2	7
9	1	4	8	2	7	3	6	5
7	9	3	5	4	2	6	1	8
4	8	5	7	1	6	2	3	9
6	2	1	9	8	3	5	7	4
1	4	7	2	6	8	9	5	3
3	6	9	1	5	4	7	8	2
8	5	2	3	7	9	1	4	6

EXPERT - 343

6	8	9	2	5	4	1	3	7
4	1	3	7	8	9	5	2	6
7	5	2	1	3	6	9	8	4
3	2	4	9	7	8	6	1	5
1	9	8	4	6	5	3	7	2
5	6	7	3	1	2	8	4	9
9	7	1	5	4	3	2	6	8
2	3	6	8	9	7	4	5	1
8	4	5	6	2	1	7	9	3

EXPERT - 344

2	9	5	1	4	3	6	8	7
1	7	6	9	5	8	4	2	3
3	4	8	2	7	6	1	5	9
6	5	2	4	8	7	9	3	1
8	1	7	3	9	2	5	4	6
9	3	4	6	1	5	2	7	8
7	8	1	5	2	9	3	6	4
5	6	9	8	3	4	7	1	2
4	2	3	7	6	1	8	9	5

EXPERT - 345

1	8	5	4	3	7	2	9	6
7	2	6	8	9	5	1	4	3
3	4	9	2	1	6	7	8	5
5	1	7	9	4	3	6	2	8
4	6	2	1	5	8	3	7	9
8	9	3	7	6	2	5	1	4
9	5	1	3	2	4	8	6	7
6	7	4	5	8	1	9	3	2
2	3	8	6	7	9	4	5	1

EXPERT - 346

3	5	4	7	2	1	8	9	6
6	7	9	8	5	4	1	2	3
2	8	1	6	3	9	4	5	7
5	1	7	2	4	6	9	3	8
4	6	2	3	9	8	5	7	1
8	9	3	1	7	5	2	6	4
1	4	5	9	6	7	3	8	2
9	3	6	4	8	2	7	1	5
7	2	8	5	1	3	6	4	9

EXPERT - 347

9	3	6	1	2	8	5	7	4
1	8	7	5	4	9	6	2	3
4	5	2	6	7	3	1	8	9
7	6	9	3	5	4	2	1	8
2	1	3	8	6	7	4	9	5
8	4	5	9	1	2	7	3	6
6	7	8	4	3	1	9	5	2
3	2	4	7	9	5	8	6	1
5	9	1	2	8	6	3	4	7

EXPERT - 348

7	1	4	2	9	8	5	6	3
6	2	5	1	7	3	8	4	9
9	3	8	5	4	6	7	1	2
5	8	3	7	6	2	1	9	4
4	7	2	3	1	9	6	8	5
1	9	6	4	8	5	3	2	7
3	5	1	6	2	4	9	7	8
2	6	9	8	3	7	4	5	1
8	4	7	9	5	1	2	3	6

EXPERT - 349

8	3	9	4	2	5	7	1	6
1	4	5	6	7	3	9	8	2
2	7	6	8	9	1	3	4	5
5	6	4	1	3	9	2	7	8
9	2	1	7	5	8	6	3	4
7	8	3	2	6	4	5	9	1
6	9	8	5	1	7	4	2	3
3	1	2	9	4	6	8	5	7
4	5	7	3	8	2	1	6	9

EXPERT - 350

8	2	7	6	1	4	5	9	3
3	5	1	7	9	2	8	6	4
4	6	9	3	5	8	1	7	2
1	7	4	2	3	5	6	8	9
9	3	6	4	8	1	2	5	7
5	8	2	9	6	7	3	4	1
7	4	3	8	2	6	9	1	5
6	9	5	1	4	3	7	2	8
2	1	8	5	7	9	4	3	6

EXPERT - 351

6	8	7	9	3	2	5	1	4
3	5	4	1	6	8	2	9	7
2	1	9	5	4	7	8	3	6
9	7	5	6	8	3	1	4	2
8	4	6	2	9	1	3	7	5
1	3	2	4	7	5	9	6	8
4	6	1	8	5	9	7	2	3
7	2	8	3	1	4	6	5	9
5	9	3	7	2	6	4	8	1

EXPERT - 352

5	4	6	9	1	2	7	3	8
7	8	3	5	4	6	9	2	1
1	9	2	8	3	7	6	4	5
6	1	4	3	5	8	2	7	9
2	7	9	4	6	1	8	5	3
8	3	5	7	2	9	1	6	4
9	5	7	2	8	3	4	1	6
3	6	8	1	7	4	5	9	2
4	2	1	6	9	5	3	8	7

EXPERT - 353

1	8	2	5	3	6	7	9	4
7	5	9	1	4	8	6	3	2
4	6	3	9	7	2	5	1	8
8	7	4	3	5	1	9	2	6
9	3	5	2	6	7	4	8	1
6	2	1	4	8	9	3	5	7
3	1	6	7	2	5	8	4	9
5	9	8	6	1	4	2	7	3
2	4	7	8	9	3	1	6	5

EXPERT - 354

9	7	4	1	8	2	5	3	6
1	8	6	4	3	5	9	7	2
3	5	2	9	6	7	4	1	8
7	9	3	8	5	4	6	2	1
4	2	8	6	1	9	7	5	3
5	6	1	2	7	3	8	9	4
6	4	7	5	2	1	3	8	9
8	1	5	3	9	6	2	4	7
2	3	9	7	4	8	1	6	5

EXPERT - 355

4	5	2	9	8	6	3	1	7
7	6	9	3	4	1	8	2	5
3	8	1	2	7	5	9	6	4
2	4	6	5	1	9	7	3	8
8	3	5	6	2	7	4	9	1
9	1	7	4	3	8	2	5	6
1	2	4	8	5	3	6	7	9
5	9	3	7	6	4	1	8	2
6	7	8	1	9	2	5	4	3

EXPERT - 356

1	3	8	6	4	5	2	9	7
5	2	6	7	1	9	8	3	4
4	9	7	8	3	2	6	5	1
7	5	2	9	8	3	1	4	6
6	8	3	4	2	1	9	7	5
9	4	1	5	7	6	3	8	2
2	6	4	3	5	8	7	1	9
3	7	9	1	6	4	5	2	8
8	1	5	2	9	7	4	6	3

EXPERT - 357

8	5	9	1	7	2	3	6	4
7	6	3	8	5	4	1	9	2
4	1	2	6	9	3	5	8	7
3	4	6	2	1	5	8	7	9
5	9	1	4	8	7	2	3	6
2	7	8	9	3	6	4	5	1
6	8	5	7	2	1	9	4	3
1	3	4	5	6	9	7	2	8
9	2	7	3	4	8	6	1	5

EXPERT - 358

7	6	3	8	2	4	9	1	5
1	9	5	3	6	7	2	4	8
4	2	8	9	5	1	7	3	6
3	8	6	4	1	2	5	9	7
5	7	2	6	9	3	4	8	1
9	4	1	5	7	8	3	6	2
2	3	9	7	8	6	1	5	4
8	1	4	2	3	5	6	7	9
6	5	7	1	4	9	8	2	3

EXPERT - 359

3	2	6	4	8	5	7	9	1
8	1	7	2	9	3	4	6	5
4	5	9	1	6	7	8	2	3
2	7	3	9	5	1	6	8	4
5	6	4	8	7	2	1	3	9
1	9	8	3	4	6	5	7	2
6	3	1	5	2	8	9	4	7
7	4	5	6	3	9	2	1	8
9	8	2	7	1	4	3	5	6

EXPERT - 360

8	3	5	2	6	4	1	9	7
6	2	9	8	1	7	5	4	3
4	1	7	5	3	9	6	2	8
1	5	2	4	7	3	9	8	6
3	6	8	9	2	1	4	7	5
7	9	4	6	8	5	3	1	2
9	7	1	3	5	8	2	6	4
2	8	3	1	4	6	7	5	9
5	4	6	7	9	2	8	3	1

EXPERT - 361

3	1	2	4	8	6	9	5	7
8	5	9	1	2	7	3	6	4
7	4	6	3	9	5	1	8	2
5	8	4	6	3	9	7	2	1
2	3	7	8	1	4	5	9	6
6	9	1	7	5	2	8	4	3
9	6	3	5	4	1	2	7	8
4	2	8	9	7	3	6	1	5
1	7	5	2	6	8	4	3	9

EXPERT - 362

2	6	8	7	3	4	5	9	1
4	7	1	9	2	5	6	8	3
9	3	5	1	6	8	4	7	2
7	4	9	2	8	1	3	6	5
6	1	3	5	9	7	2	4	8
8	5	2	6	4	3	9	1	7
5	8	6	3	1	9	7	2	4
1	2	7	4	5	6	8	3	9
3	9	4	8	7	2	1	5	6

EXPERT - 363

4	7	5	3	1	9	8	2	6
2	1	3	8	6	7	9	4	5
9	6	8	4	2	5	7	1	3
7	5	6	9	8	1	2	3	4
8	2	9	5	4	3	1	6	7
3	4	1	2	7	6	5	9	8
1	8	2	6	5	4	3	7	9
6	9	7	1	3	8	4	5	2
5	3	4	7	9	2	6	8	1

EXPERT - 364

1	9	4	5	6	7	3	8	2
5	2	3	4	9	8	6	1	7
8	6	7	1	2	3	5	4	9
4	8	9	7	3	5	1	2	6
3	1	6	9	4	2	8	7	5
2	7	5	6	8	1	4	9	3
7	4	1	3	5	9	2	6	8
6	3	2	8	7	4	9	5	1
9	5	8	2	1	6	7	3	4

EXPERT - 365

9	3	7	4	6	1	2	5	8
5	4	2	9	7	8	6	3	1
8	6	1	2	3	5	7	9	4
4	7	3	1	9	6	5	8	2
1	8	5	7	4	2	9	6	3
2	9	6	5	8	3	1	4	7
7	5	8	3	1	9	4	2	6
3	1	9	6	2	4	8	7	5
6	2	4	8	5	7	3	1	9

EXPERT - 366

6	2	1	9	7	3	8	5	4
5	3	8	1	6	4	2	9	7
4	7	9	8	5	2	3	1	6
3	6	7	2	9	5	1	4	8
1	5	2	4	8	7	9	6	3
8	9	4	6	3	1	7	2	5
7	1	3	5	4	9	6	8	2
2	4	6	7	1	8	5	3	9
9	8	5	3	2	6	4	7	1

EXPERT - 367

1	2	8	4	6	3	5	9	7
6	5	9	8	2	7	1	3	4
7	3	4	9	1	5	6	8	2
9	4	5	3	8	1	7	2	6
2	6	1	7	9	4	8	5	3
3	8	7	2	5	6	9	4	1
4	1	3	5	7	8	2	6	9
8	9	6	1	4	2	3	7	5
5	7	2	6	3	9	4	1	8

EXPERT - 368

5	8	1	4	9	6	3	7	2
6	4	3	8	2	7	5	9	1
7	2	9	1	5	3	8	4	6
3	7	4	2	8	9	6	1	5
9	6	8	5	7	1	2	3	4
1	5	2	6	3	4	9	8	7
4	3	7	9	6	5	1	2	8
8	1	5	3	4	2	7	6	9
2	9	6	7	1	8	4	5	3

EXPERT - 369

7	6	9	1	5	8	3	4	2
8	5	4	3	2	7	6	1	9
2	3	1	6	4	9	8	7	5
5	9	8	2	6	1	4	3	7
6	4	3	7	8	5	2	9	1
1	2	7	4	9	3	5	6	8
3	1	2	8	7	6	9	5	4
4	7	5	9	3	2	1	8	6
9	8	6	5	1	4	7	2	3

EXPERT - 370

1	9	6	7	5	2	3	4	8
5	2	8	3	9	4	1	7	6
3	7	4	1	6	8	5	2	9
6	1	9	8	4	5	2	3	7
2	4	3	6	1	7	9	8	5
8	5	7	9	2	3	6	1	4
7	6	1	4	3	9	8	5	2
9	8	5	2	7	1	4	6	3
4	3	2	5	8	6	7	9	1

EXPERT - 371

4	8	3	1	5	9	2	6	7
7	1	2	8	6	4	3	9	5
6	5	9	3	2	7	8	4	1
2	6	8	4	1	3	5	7	9
5	4	7	2	9	8	1	3	6
3	9	1	5	7	6	4	2	8
8	3	6	9	4	5	7	1	2
9	2	5	7	3	1	6	8	4
1	7	4	6	8	2	9	5	3

EXPERT - 372

6	3	5	7	8	9	4	1	2
4	2	9	6	1	3	7	8	5
1	8	7	2	5	4	3	9	6
3	7	6	8	2	5	9	4	1
8	9	2	3	4	1	5	6	7
5	4	1	9	6	7	8	2	3
7	1	8	5	9	2	6	3	4
2	6	3	4	7	8	1	5	9
9	5	4	1	3	6	2	7	8

EXPERT - 373

9	2	7	5	6	4	1	3	8
4	8	3	1	7	2	9	6	5
6	1	5	9	3	8	4	2	7
5	4	6	3	2	1	7	8	9
2	7	1	8	9	5	6	4	3
8	3	9	7	4	6	5	1	2
1	5	2	6	8	9	3	7	4
3	6	8	4	5	7	2	9	1
7	9	4	2	1	3	8	5	6

EXPERT - 374

4	2	5	7	1	6	9	3	8
9	1	8	4	3	5	2	6	7
6	3	7	8	2	9	4	1	5
1	6	9	5	8	3	7	4	2
3	8	2	6	7	4	5	9	1
5	7	4	1	9	2	3	8	6
2	9	1	3	6	7	8	5	4
8	4	3	2	5	1	6	7	9
7	5	6	9	4	8	1	2	3

EXPERT - 375

5	7	8	6	3	4	9	2	1
2	1	4	5	7	9	6	3	8
3	6	9	2	8	1	5	7	4
4	8	6	3	1	2	7	9	5
7	9	2	8	6	5	4	1	3
1	3	5	4	9	7	2	8	6
8	5	1	9	2	6	3	4	7
9	4	7	1	5	3	8	6	2
6	2	3	7	4	8	1	5	9

EXPERT - 376

4	6	3	8	1	7	5	9	2
2	1	8	6	5	9	3	4	7
9	5	7	3	2	4	8	1	6
3	2	1	9	4	8	7	6	5
6	4	9	5	7	2	1	3	8
8	7	5	1	3	6	9	2	4
5	3	2	7	6	1	4	8	9
7	8	4	2	9	3	6	5	1
1	9	6	4	8	5	2	7	3

EXPERT - 377

7	3	4	9	5	6	8	2	1
9	1	6	4	8	2	3	5	7
5	8	2	1	7	3	6	9	4
6	7	3	8	1	5	2	4	9
8	4	5	2	9	7	1	6	3
1	2	9	6	3	4	5	7	8
4	9	8	5	6	1	7	3	2
2	6	7	3	4	8	9	1	5
3	5	1	7	2	9	4	8	6

EXPERT - 378

4	6	5	9	8	7	1	3	2
7	8	2	3	4	1	5	9	6
1	3	9	5	6	2	8	4	7
9	5	1	4	7	6	2	8	3
3	2	7	8	5	9	4	6	1
8	4	6	1	2	3	9	7	5
5	9	3	7	1	8	6	2	4
6	1	8	2	3	4	7	5	9
2	7	4	6	9	5	3	1	8

EXPERT - 379

7	8	9	3	2	5	4	1	6
2	5	6	9	4	1	7	3	8
1	3	4	6	8	7	5	9	2
8	7	3	5	1	9	6	2	4
6	2	1	7	3	4	9	8	5
4	9	5	8	6	2	1	7	3
3	4	2	1	7	6	8	5	9
9	6	7	2	5	8	3	4	1
5	1	8	4	9	3	2	6	7

EXPERT - 380

7	3	5	2	4	9	6	1	8
9	4	2	1	8	6	3	7	5
1	6	8	7	5	3	9	2	4
6	5	3	9	2	8	1	4	7
2	8	9	4	7	1	5	6	3
4	1	7	6	3	5	8	9	2
8	9	4	5	6	2	7	3	1
3	2	1	8	9	7	4	5	6
5	7	6	3	1	4	2	8	9

EXPERT - 381

2	6	1	5	4	9	8	7	3
7	5	3	8	1	6	2	9	4
9	8	4	2	7	3	5	6	1
3	4	8	9	5	1	6	2	7
1	7	5	6	2	8	3	4	9
6	9	2	4	3	7	1	8	5
4	1	7	3	8	2	9	5	6
8	3	6	7	9	5	4	1	2
5	2	9	1	6	4	7	3	8

EXPERT - 382

7	6	1	3	2	5	4	8	9
9	4	8	7	1	6	3	5	2
3	5	2	4	8	9	6	7	1
5	7	9	6	4	1	2	3	8
2	3	4	9	7	8	1	6	5
1	8	6	5	3	2	9	4	7
6	2	3	8	9	7	5	1	4
8	9	5	1	6	4	7	2	3
4	1	7	2	5	3	8	9	6

EXPERT - 383

4	3	6	9	7	1	5	8	2
9	7	8	2	6	5	4	1	3
2	1	5	8	4	3	7	9	6
5	4	1	7	2	9	6	3	8
6	2	3	4	5	8	9	7	1
7	8	9	3	1	6	2	4	5
1	5	7	6	8	4	3	2	9
8	9	2	5	3	7	1	6	4
3	6	4	1	9	2	8	5	7

EXPERT - 384

4	7	8	6	9	1	3	5	2
6	1	2	5	8	3	4	7	9
3	5	9	4	2	7	1	6	8
7	9	1	2	3	4	5	8	6
8	3	6	7	5	9	2	1	4
5	2	4	8	1	6	9	3	7
9	6	7	3	4	5	8	2	1
1	8	5	9	7	2	6	4	3
2	4	3	1	6	8	7	9	5

EXPERT - 385

2	7	9	6	5	3	4	1	8
8	4	3	1	7	2	6	9	5
1	5	6	9	8	4	3	2	7
6	1	5	4	3	7	2	8	9
3	8	7	2	9	1	5	6	4
9	2	4	5	6	8	1	7	3
4	3	2	7	1	9	8	5	6
7	6	8	3	2	5	9	4	1
5	9	1	8	4	6	7	3	2

EXPERT - 386

3	6	4	9	1	8	2	5	7
7	5	2	3	6	4	8	1	9
8	9	1	7	2	5	3	4	6
1	3	6	4	9	2	5	7	8
4	2	9	8	5	7	6	3	1
5	8	7	1	3	6	9	2	4
9	4	5	6	7	3	1	8	2
2	1	8	5	4	9	7	6	3
6	7	3	2	8	1	4	9	5

EXPERT - 387

6	4	9	1	3	2	8	5	7
7	1	2	8	5	6	4	3	9
8	5	3	7	4	9	6	1	2
4	3	1	5	9	7	2	6	8
2	9	8	3	6	1	7	4	5
5	7	6	2	8	4	3	9	1
1	2	5	6	7	3	9	8	4
9	6	7	4	1	8	5	2	3
3	8	4	9	2	5	1	7	6

EXPERT - 388

3	5	4	2	6	1	9	8	7
9	2	1	7	3	8	5	6	4
7	6	8	9	4	5	1	3	2
5	9	3	8	2	7	4	1	6
2	4	7	6	1	3	8	9	5
8	1	6	5	9	4	2	7	3
6	3	9	1	5	2	7	4	8
4	7	5	3	8	9	6	2	1
1	8	2	4	7	6	3	5	9

EXPERT - 389

1	2	7	8	6	4	5	9	3
6	8	3	9	2	5	4	1	7
5	4	9	3	7	1	6	2	8
7	5	6	2	1	8	9	3	4
8	1	4	6	9	3	2	7	5
3	9	2	4	5	7	8	6	1
9	3	8	7	4	2	1	5	6
4	6	5	1	3	9	7	8	2
2	7	1	5	8	6	3	4	9

EXPERT - 390

1	2	8	3	9	7	5	4	6
9	6	7	5	2	4	3	1	8
4	5	3	6	1	8	7	2	9
2	3	1	7	4	9	8	6	5
6	4	9	8	5	2	1	7	3
8	7	5	1	3	6	4	9	2
5	8	4	9	6	1	2	3	7
7	1	6	2	8	3	9	5	4
3	9	2	4	7	5	6	8	1

EXPERT - 391

9	5	2	1	6	8	7	3	4
1	3	4	9	5	7	8	6	2
8	7	6	4	2	3	5	1	9
4	1	3	7	8	2	6	9	5
7	2	8	5	9	6	3	4	1
6	9	5	3	4	1	2	7	8
5	6	1	8	3	9	4	2	7
2	4	7	6	1	5	9	8	3
3	8	9	2	7	4	1	5	6

EXPERT - 392

3	9	1	8	2	6	5	4	7
5	6	7	4	9	1	3	2	8
8	4	2	5	3	7	9	6	1
9	5	3	2	6	8	1	7	4
7	2	6	1	4	5	8	9	3
4	1	8	9	7	3	6	5	2
1	7	4	3	5	9	2	8	6
6	8	9	7	1	2	4	3	5
2	3	5	6	8	4	7	1	9

EXPERT - 393

6	2	5	8	3	4	1	9	7
3	8	7	9	5	1	6	4	2
9	1	4	2	7	6	3	5	8
7	5	3	4	8	9	2	6	1
8	4	1	5	6	2	7	3	9
2	6	9	7	1	3	4	8	5
1	9	6	3	2	5	8	7	4
5	7	2	6	4	8	9	1	3
4	3	8	1	9	7	5	2	6

EXPERT - 394

6	1	7	4	8	2	9	5	3
9	8	4	3	6	5	1	2	7
2	3	5	7	1	9	8	4	6
4	6	9	1	2	3	7	8	5
5	2	3	9	7	8	4	6	1
1	7	8	6	5	4	2	3	9
3	4	2	5	9	7	6	1	8
7	5	1	8	4	6	3	9	2
8	9	6	2	3	1	5	7	4

EXPERT - 395

9	3	6	7	8	1	2	5	4
1	5	7	3	2	4	8	6	9
4	8	2	6	9	5	1	3	7
6	9	3	1	4	2	5	7	8
8	4	1	5	7	9	6	2	3
7	2	5	8	6	3	4	9	1
2	7	4	9	1	6	3	8	5
5	6	8	4	3	7	9	1	2
3	1	9	2	5	8	7	4	6

EXPERT - 396

3	5	8	2	6	9	4	7	1
4	2	1	8	7	5	3	6	9
9	7	6	3	1	4	2	5	8
1	6	7	9	4	3	5	8	2
5	4	2	7	8	6	9	1	3
8	3	9	5	2	1	6	4	7
7	8	4	6	3	2	1	9	5
6	9	3	1	5	8	7	2	4
2	1	5	4	9	7	8	3	6

EXPERT - 397

5	3	9	1	6	7	2	4	8
7	4	2	3	5	8	1	9	6
8	1	6	2	4	9	7	5	3
2	5	3	8	9	1	4	6	7
6	8	4	5	7	3	9	2	1
9	7	1	6	2	4	8	3	5
3	9	5	7	8	2	6	1	4
1	2	8	4	3	6	5	7	9
4	6	7	9	1	5	3	8	2

EXPERT - 398

6	8	4	5	2	3	9	1	7
7	3	9	1	4	6	8	2	5
5	1	2	7	9	8	3	6	4
2	9	8	3	1	5	4	7	6
1	7	3	4	6	9	5	8	2
4	6	5	8	7	2	1	9	3
8	2	7	9	5	4	6	3	1
9	4	1	6	3	7	2	5	8
3	5	6	2	8	1	7	4	9

EXPERT - 399

9	7	6	3	1	4	2	5	8
4	1	2	6	5	8	3	7	9
5	3	8	9	7	2	1	6	4
8	9	7	1	2	3	5	4	6
2	6	1	7	4	5	8	9	3
3	5	4	8	6	9	7	2	1
7	4	9	5	8	1	6	3	2
6	8	3	2	9	7	4	1	5
1	2	5	4	3	6	9	8	7

EXPERT - 400

9	7	1	3	6	2	5	4	8
2	6	3	4	8	5	1	9	7
8	4	5	7	1	9	6	3	2
4	8	6	2	5	7	9	1	3
7	5	9	1	3	6	2	8	4
1	3	2	8	9	4	7	5	6
6	1	8	5	7	3	4	2	9
3	2	7	9	4	1	8	6	5
5	9	4	6	2	8	3	7	1

EXPERT - 401

9	3	1	8	2	6	5	7	4
8	7	6	4	3	5	2	9	1
4	2	5	9	7	1	6	8	3
6	9	7	1	8	2	3	4	5
2	5	3	6	9	4	7	1	8
1	4	8	3	5	7	9	2	6
7	8	4	2	6	3	1	5	9
5	6	9	7	1	8	4	3	2
3	1	2	5	4	9	8	6	7

EXPERT - 402

8	9	2	6	5	3	1	7	4
3	5	4	1	7	9	6	8	2
7	6	1	2	8	4	5	9	3
2	8	9	7	4	5	3	1	6
4	3	6	9	2	1	8	5	7
5	1	7	3	6	8	2	4	9
6	4	5	8	3	7	9	2	1
9	2	8	4	1	6	7	3	5
1	7	3	5	9	2	4	6	8

EXPERT - 403

9	2	1	5	4	7	3	6	8
4	6	3	9	1	8	2	5	7
8	5	7	2	3	6	9	4	1
6	3	5	1	7	9	4	8	2
1	7	9	4	8	2	5	3	6
2	8	4	3	6	5	7	1	9
3	1	6	7	9	4	8	2	5
5	9	8	6	2	3	1	7	4
7	4	2	8	5	1	6	9	3

EXPERT - 404

2	3	8	9	7	1	6	5	4
7	1	6	4	8	5	2	3	9
9	5	4	2	3	6	7	1	8
8	9	3	5	6	4	1	2	7
6	7	5	1	2	8	9	4	3
4	2	1	3	9	7	5	8	6
3	4	7	6	1	2	8	9	5
5	6	2	8	4	9	3	7	1
1	8	9	7	5	3	4	6	2

EXPERT - 405

5	4	8	9	2	7	6	1	3
3	7	1	4	5	6	8	2	9
2	9	6	1	8	3	5	7	4
4	8	7	3	6	1	2	9	5
9	2	5	8	7	4	1	3	6
1	6	3	2	9	5	4	8	7
8	3	2	6	4	9	7	5	1
7	1	4	5	3	2	9	6	8
6	5	9	7	1	8	3	4	2

EXPERT - 406

3	6	8	4	9	7	1	5	2
1	2	7	8	6	5	3	4	9
9	5	4	1	2	3	8	7	6
2	3	6	7	8	9	4	1	5
7	8	1	5	4	6	2	9	3
5	4	9	3	1	2	6	8	7
4	7	5	2	3	1	9	6	8
8	9	2	6	7	4	5	3	1
6	1	3	9	5	8	7	2	4

EXPERT - 407

5	3	7	1	8	6	2	9	4
2	6	1	4	9	7	3	8	5
4	8	9	2	3	5	6	7	1
6	1	4	8	2	9	7	5	3
8	7	5	3	6	4	9	1	2
3	9	2	7	5	1	8	4	6
7	4	3	6	1	8	5	2	9
9	2	8	5	4	3	1	6	7
1	5	6	9	7	2	4	3	8

EXPERT - 408

2	6	8	7	3	4	5	9	1
4	7	1	8	9	5	2	3	6
5	9	3	6	2	1	7	4	8
1	5	6	9	4	8	3	2	7
9	8	7	3	6	2	1	5	4
3	4	2	1	5	7	6	8	9
7	2	5	4	8	6	9	1	3
6	3	4	5	1	9	8	7	2
8	1	9	2	7	3	4	6	5

EXPERT - 409

8	9	2	3	4	6	1	5	7
5	4	6	8	7	1	9	3	2
3	7	1	2	9	5	8	4	6
1	3	4	6	8	9	2	7	5
6	5	9	7	2	3	4	1	8
7	2	8	5	1	4	3	6	9
9	1	7	4	6	2	5	8	3
2	8	5	1	3	7	6	9	4
4	6	3	9	5	8	7	2	1

EXPERT - 410

4	6	5	8	3	9	7	1	2
7	9	1	4	2	5	3	6	8
3	8	2	1	6	7	4	5	9
6	4	8	7	1	3	2	9	5
5	2	7	9	8	4	1	3	6
9	1	3	2	5	6	8	4	7
1	5	6	3	7	8	9	2	4
2	7	9	6	4	1	5	8	3
8	3	4	5	9	2	6	7	1

EXPERT - 411

1	5	9	6	3	7	2	4	8
4	6	8	9	2	5	7	3	1
2	7	3	4	8	1	5	9	6
5	9	4	7	1	3	8	6	2
8	3	6	5	4	2	1	7	9
7	2	1	8	9	6	4	5	3
6	1	7	2	5	9	3	8	4
9	8	2	3	7	4	6	1	5
3	4	5	1	6	8	9	2	7

EXPERT - 412

8	1	6	7	4	2	9	5	3
2	4	3	5	1	9	8	6	7
9	5	7	3	6	8	2	4	1
1	3	8	6	7	4	5	9	2
5	9	4	1	2	3	7	8	6
6	7	2	8	9	5	3	1	4
3	2	9	4	8	6	1	7	5
7	6	5	9	3	1	4	2	8
4	8	1	2	5	7	6	3	9

EXPERT - 413

8	7	6	9	4	1	5	2	3
1	2	9	3	6	5	8	4	7
4	5	3	8	2	7	1	6	9
7	9	5	6	1	3	4	8	2
6	1	4	2	7	8	9	3	5
3	8	2	5	9	4	7	1	6
9	3	7	4	8	2	6	5	1
2	4	1	7	5	6	3	9	8
5	6	8	1	3	9	2	7	4

EXPERT - 414

6	7	2	1	9	5	3	4	8
3	8	9	2	6	4	5	1	7
1	5	4	3	7	8	2	9	6
5	2	3	7	8	9	1	6	4
7	4	6	5	3	1	9	8	2
8	9	1	6	4	2	7	5	3
9	6	5	8	2	3	4	7	1
4	3	7	9	1	6	8	2	5
2	1	8	4	5	7	6	3	9

EXPERT - 415

9	2	1	4	8	5	6	7	3
5	7	6	9	1	3	4	2	8
4	8	3	2	6	7	5	9	1
7	6	8	1	3	2	9	5	4
3	9	4	6	5	8	7	1	2
2	1	5	7	9	4	8	3	6
1	8	7	3	4	9	2	6	5
4	3	9	5	2	6	1	8	7
6	5	2	8	7	1	3	4	9

EXPERT - 416

6	4	3	5	8	1	2	7	9
7	9	8	3	2	6	1	4	5
2	5	1	9	7	4	8	6	3
8	3	2	6	1	9	7	5	4
9	1	5	4	3	7	6	2	8
4	7	6	2	5	8	9	3	1
1	2	7	8	4	5	3	9	6
5	8	9	7	6	3	4	1	2
3	6	4	1	9	2	5	8	7

EXPERT - 417

1	2	4	8	5	6	3	7	9
8	3	7	2	9	1	6	4	5
5	6	9	3	4	7	1	8	2
3	9	8	1	6	4	5	2	7
2	4	6	7	3	5	9	1	8
7	5	1	9	8	2	4	3	6
6	7	2	5	1	3	8	9	4
9	1	5	4	7	8	2	6	3
4	8	3	6	2	9	7	5	1

EXPERT - 418

9	2	5	7	4	1	6	3	8
4	3	1	8	2	6	9	7	5
7	6	8	5	9	3	1	4	2
1	4	7	2	6	5	3	8	9
5	9	2	3	1	8	7	6	4
3	8	6	4	7	9	5	2	1
6	7	4	1	5	2	8	9	3
2	1	3	9	8	7	4	5	6
8	5	9	6	3	4	2	1	7

EXPERT - 419

1	7	8	4	6	5	9	3	2
6	4	2	3	9	8	1	7	5
9	3	5	2	1	7	4	8	6
7	2	9	5	8	6	3	1	4
4	5	6	7	3	1	2	9	8
3	8	1	9	4	2	6	5	7
2	9	7	6	5	3	8	4	1
8	6	3	1	7	4	5	2	9
5	1	4	8	2	9	7	6	3

EXPERT - 420

7	3	5	9	6	2	8	4	1
6	9	4	1	8	5	7	2	3
1	8	2	4	3	7	9	6	5
4	2	1	3	5	8	6	9	7
8	5	6	7	2	9	1	3	4
3	7	9	6	4	1	2	5	8
5	6	7	8	9	3	4	1	2
2	4	8	5	1	6	3	7	9
9	1	3	2	7	4	5	8	6

EXPERT - 421

4	2	7	9	8	3	5	1	6
5	6	8	1	7	4	9	3	2
9	1	3	5	2	6	7	8	4
2	3	9	8	4	7	1	6	5
8	7	5	2	6	1	3	4	9
6	4	1	3	9	5	8	2	7
7	8	4	6	3	9	2	5	1
3	5	6	7	1	2	4	9	8
1	9	2	4	5	8	6	7	3

EXPERT - 422

8	1	5	4	9	6	2	7	3
3	9	2	7	5	8	1	6	4
6	7	4	1	3	2	9	8	5
2	8	9	6	7	4	3	5	1
5	3	7	2	1	9	8	4	6
4	6	1	5	8	3	7	9	2
9	2	3	8	6	5	4	1	7
7	4	6	9	2	1	5	3	8
1	5	8	3	4	7	6	2	9

EXPERT - 423

7	5	4	2	9	8	3	6	1
6	9	3	7	1	5	2	8	4
2	8	1	3	4	6	7	9	5
5	6	2	1	3	7	9	4	8
9	3	8	5	2	4	6	1	7
1	4	7	8	6	9	5	3	2
3	7	5	9	8	1	4	2	6
4	1	9	6	7	2	8	5	3
8	2	6	4	5	3	1	7	9

EXPERT - 424

6	2	5	1	9	3	8	4	7
7	8	9	6	4	2	3	5	1
3	1	4	8	7	5	9	2	6
9	4	3	2	1	7	5	6	8
2	7	8	4	5	6	1	9	3
5	6	1	3	8	9	2	7	4
8	9	7	5	3	4	6	1	2
4	3	2	9	6	1	7	8	5
1	5	6	7	2	8	4	3	9

EXPERT - 425

2	1	6	3	9	4	5	7	8
8	3	7	1	6	5	4	9	2
5	4	9	8	2	7	1	6	3
4	9	3	2	7	1	8	5	6
6	2	1	4	5	8	9	3	7
7	8	5	6	3	9	2	1	4
1	6	8	5	4	3	7	2	9
3	7	4	9	1	2	6	8	5
9	5	2	7	8	6	3	4	1

EXPERT - 426

6	2	9	4	1	7	3	8	5
3	7	4	9	8	5	1	6	2
1	8	5	6	3	2	9	7	4
5	1	6	3	4	9	7	2	8
8	4	2	7	6	1	5	9	3
7	9	3	5	2	8	6	4	1
4	5	8	1	9	6	2	3	7
2	6	7	8	5	3	4	1	9
9	3	1	2	7	4	8	5	6

EXPERT - 427

9	6	2	1	7	3	8	5	4
1	8	4	5	6	2	7	9	3
3	5	7	9	4	8	1	2	6
2	3	1	6	8	9	4	7	5
4	7	6	3	5	1	2	8	9
8	9	5	7	2	4	6	3	1
7	1	8	4	9	5	3	6	2
6	4	9	2	3	7	5	1	8
5	2	3	8	1	6	9	4	7

EXPERT - 428

3	7	8	1	5	2	6	9	4
9	6	5	7	4	8	1	2	3
2	1	4	9	6	3	5	7	8
1	8	3	5	9	7	2	4	6
6	2	9	4	3	1	8	5	7
4	5	7	2	8	6	9	3	1
7	4	2	6	1	5	3	8	9
5	3	1	8	7	9	4	6	2
8	9	6	3	2	4	7	1	5

EXPERT - 429

6	5	2	8	4	7	1	9	3
3	7	1	6	5	9	4	2	8
4	9	8	3	1	2	6	7	5
1	2	6	5	3	4	9	8	7
8	3	9	1	7	6	2	5	4
7	4	5	2	9	8	3	1	6
9	6	3	7	2	5	8	4	1
2	8	7	4	6	1	5	3	9
5	1	4	9	8	3	7	6	2

EXPERT - 430

4	8	1	6	5	9	7	3	2
2	6	3	1	8	7	5	4	9
5	7	9	4	3	2	1	8	6
7	4	8	5	9	6	2	1	3
9	3	5	2	4	1	6	7	8
1	2	6	8	7	3	9	5	4
8	5	2	9	1	4	3	6	7
6	1	7	3	2	8	4	9	5
3	9	4	7	6	5	8	2	1

EXPERT - 431

6	7	8	9	4	3	1	2	5
2	5	4	1	6	8	7	9	3
3	9	1	5	7	2	4	6	8
1	8	5	4	3	9	2	7	6
7	4	6	8	2	5	3	1	9
9	2	3	6	1	7	5	8	4
8	1	9	2	5	4	6	3	7
5	3	2	7	8	6	9	4	1
4	6	7	3	9	1	8	5	2

EXPERT - 432

5	2	6	1	7	8	9	3	4
8	9	3	4	2	5	7	6	1
7	4	1	9	6	3	5	2	8
9	7	8	2	4	6	1	5	3
4	3	5	8	1	7	2	9	6
6	1	2	5	3	9	8	4	7
1	6	4	7	9	2	3	8	5
3	5	9	6	8	1	4	7	2
2	8	7	3	5	4	6	1	9

EXPERT - 433

7	9	4	8	2	6	3	1	5
3	5	6	4	9	1	8	2	7
8	1	2	5	3	7	9	6	4
2	4	5	7	8	9	6	3	1
1	6	3	2	4	5	7	8	9
9	7	8	1	6	3	5	4	2
5	8	1	6	7	2	4	9	3
6	2	9	3	5	4	1	7	8
4	3	7	9	1	8	2	5	6

EXPERT - 434

7	3	8	2	4	9	6	5	1
2	1	9	5	7	6	8	3	4
5	4	6	3	1	8	9	2	7
9	8	5	1	6	3	4	7	2
4	2	1	7	9	5	3	8	6
3	6	7	4	8	2	1	9	5
1	5	2	9	3	4	7	6	8
6	9	4	8	2	7	5	1	3
8	7	3	6	5	1	2	4	9

EXPERT - 435

3	7	6	4	2	1	8	9	5
4	2	5	9	3	8	6	7	1
1	8	9	7	5	6	3	2	4
8	6	1	5	7	3	9	4	2
7	4	2	1	6	9	5	8	3
5	9	3	2	8	4	7	1	6
9	1	7	3	4	5	2	6	8
6	3	4	8	9	2	1	5	7
2	5	8	6	1	7	4	3	9

EXPERT - 436

4	6	2	1	8	5	7	9	3
5	1	9	3	2	7	6	4	8
7	3	8	6	9	4	2	5	1
2	9	7	4	6	3	8	1	5
3	5	6	9	1	8	4	2	7
1	8	4	5	7	2	3	6	9
9	2	1	7	3	6	5	8	4
6	7	5	8	4	1	9	3	2
8	4	3	2	5	9	1	7	6

EXPERT - 437

3	7	9	5	6	1	8	4	2
6	4	2	7	8	3	1	9	5
8	1	5	4	9	2	7	3	6
1	6	8	2	7	4	9	5	3
7	9	3	8	5	6	2	1	4
5	2	4	3	1	9	6	8	7
2	8	1	6	3	5	4	7	9
4	5	7	9	2	8	3	6	1
9	3	6	1	4	7	5	2	8

EXPERT - 438

3	2	1	4	7	5	8	9	6
5	4	6	8	3	9	1	7	2
9	7	8	2	6	1	3	4	5
2	5	3	9	4	8	6	1	7
7	6	9	1	2	3	5	8	4
1	8	4	6	5	7	9	2	3
8	3	5	7	1	4	2	6	9
6	1	7	3	9	2	4	5	8
4	9	2	5	8	6	7	3	1

EXPERT - 439

3	1	8	4	5	6	9	2	7
2	4	5	7	1	9	6	8	3
7	6	9	8	3	2	1	4	5
6	9	4	3	8	1	7	5	2
5	7	1	9	2	4	8	3	6
8	2	3	5	6	7	4	9	1
4	3	6	1	9	5	2	7	8
9	5	2	6	7	8	3	1	4
1	8	7	2	4	3	5	6	9

EXPERT - 440

8	9	1	6	7	4	5	3	2
4	3	5	1	2	9	6	8	7
6	7	2	3	8	5	1	4	9
3	2	6	4	9	8	7	5	1
5	8	9	7	6	1	4	2	3
1	4	7	5	3	2	9	6	8
2	6	8	9	5	7	3	1	4
9	1	3	8	4	6	2	7	5
7	5	4	2	1	3	8	9	6

EXPERT - 441

6	3	2	9	8	7	1	4	5
1	9	7	4	5	6	2	3	8
5	4	8	1	3	2	7	6	9
4	2	9	7	1	3	5	8	6
3	1	5	6	4	8	9	7	2
7	8	6	2	9	5	3	1	4
8	7	4	3	2	9	6	5	1
2	5	3	8	6	1	4	9	7
9	6	1	5	7	4	8	2	3

EXPERT - 442

3	8	5	7	2	1	6	4	9
9	1	2	5	6	4	7	8	3
7	4	6	8	3	9	1	5	2
1	9	3	6	4	5	8	2	7
5	7	8	3	1	2	4	9	6
6	2	4	9	7	8	5	3	1
2	5	1	4	9	6	3	7	8
8	3	9	1	5	7	2	6	4
4	6	7	2	8	3	9	1	5

EXPERT - 443

6	5	8	9	3	1	7	4	2
7	4	2	8	5	6	9	1	3
1	9	3	4	2	7	5	6	8
4	3	7	2	8	5	1	9	6
5	1	6	3	7	9	2	8	4
2	8	9	6	1	4	3	7	5
3	7	5	1	6	8	4	2	9
8	2	4	7	9	3	6	5	1
9	6	1	5	4	2	8	3	7

EXPERT - 444

3	1	8	2	4	5	7	6	9
9	2	5	7	3	6	4	8	1
4	7	6	8	9	1	5	3	2
5	9	7	6	8	4	1	2	3
1	6	4	3	5	2	9	7	8
2	8	3	1	7	9	6	4	5
6	3	2	9	1	7	8	5	4
7	4	9	5	2	8	3	1	6
8	5	1	4	6	3	2	9	7

EXPERT - 445

1	8	4	7	3	9	2	5	6
2	7	3	6	5	4	8	1	9
9	5	6	8	1	2	4	7	3
4	9	8	2	6	5	7	3	1
6	3	7	4	8	1	9	2	5
5	1	2	9	7	3	6	4	8
3	4	9	5	2	6	1	8	7
7	6	5	1	4	8	3	9	2
8	2	1	3	9	7	5	6	4

EXPERT - 446

5	1	8	7	4	9	6	2	3
7	3	9	6	2	5	8	1	4
4	6	2	8	3	1	9	7	5
1	4	5	9	6	3	7	8	2
6	9	7	5	8	2	4	3	1
8	2	3	4	1	7	5	6	9
2	7	4	1	5	8	3	9	6
9	5	1	3	7	6	2	4	8
3	8	6	2	9	4	1	5	7

EXPERT - 447

8	9	5	4	1	6	3	7	2
2	1	3	5	8	7	4	6	9
4	7	6	9	2	3	8	1	5
3	6	1	2	9	4	5	8	7
7	5	2	6	3	8	9	4	1
9	8	4	7	5	1	2	3	6
6	3	9	1	4	5	7	2	8
1	2	8	3	7	9	6	5	4
5	4	7	8	6	2	1	9	3

EXPERT - 448

1	6	9	7	5	3	4	8	2
7	4	5	8	2	1	9	3	6
3	2	8	4	9	6	7	1	5
9	1	7	5	3	4	6	2	8
4	8	6	2	1	7	5	9	3
5	3	2	9	6	8	1	4	7
2	5	4	6	8	9	3	7	1
8	7	3	1	4	5	2	6	9
6	9	1	3	7	2	8	5	4

EXPERT - 449

1	7	4	3	2	5	8	6	9
3	9	6	1	8	7	5	2	4
5	8	2	4	6	9	7	1	3
2	1	5	9	3	8	6	4	7
8	4	3	7	1	6	2	9	5
9	6	7	5	4	2	1	3	8
6	2	9	8	5	3	4	7	1
4	3	8	6	7	1	9	5	2
7	5	1	2	9	4	3	8	6

EXPERT - 450

5	6	2	7	4	8	9	3	1
3	8	1	9	5	2	6	7	4
7	9	4	6	3	1	2	8	5
4	3	6	2	8	9	5	1	7
9	1	8	5	6	7	3	4	2
2	5	7	3	1	4	8	6	9
6	7	3	4	2	5	1	9	8
8	4	5	1	9	6	7	2	3
1	2	9	8	7	3	4	5	6

EXPERT - 451

5	3	2	1	8	9	6	7	4
6	7	4	3	2	5	9	1	8
1	8	9	4	7	6	5	3	2
4	2	5	8	6	3	7	9	1
8	9	1	7	5	2	4	6	3
3	6	7	9	1	4	2	8	5
9	1	3	2	4	7	8	5	6
7	4	6	5	3	8	1	2	9
2	5	8	6	9	1	3	4	7

EXPERT - 452

6	2	3	9	5	8	4	1	7
9	7	5	1	4	6	2	8	3
4	8	1	3	2	7	6	5	9
1	9	2	5	6	4	7	3	8
7	5	4	8	3	2	9	6	1
8	3	6	7	1	9	5	2	4
5	4	8	2	9	3	1	7	6
3	1	9	6	7	5	8	4	2
2	6	7	4	8	1	3	9	5

EXPERT - 453

7	1	8	3	9	2	4	5	6
4	5	9	7	6	1	8	2	3
2	6	3	8	5	4	9	7	1
9	8	5	6	1	3	7	2	4
1	7	4	2	8	5	6	3	9
6	3	2	9	4	7	5	1	8
8	4	1	5	7	6	3	9	2
5	2	6	1	3	9	8	4	7
3	9	7	4	2	8	1	6	5

EXPERT - 454

4	1	3	2	6	7	8	9	5
5	8	7	1	9	4	6	3	2
9	6	2	5	8	3	4	7	1
6	2	9	3	7	5	1	4	8
1	4	5	8	2	9	3	6	7
7	3	8	4	1	6	2	5	9
3	5	1	9	4	8	7	2	6
2	7	4	6	5	1	9	8	3
8	9	6	7	3	2	5	1	4

EXPERT - 455

6	2	9	3	7	4	8	1	5
8	4	5	9	2	1	7	3	6
3	7	1	6	5	8	9	4	2
4	6	3	8	9	7	5	2	1
7	1	8	2	4	5	3	6	9
9	5	2	1	3	6	4	8	7
2	8	7	5	1	3	6	9	4
1	3	4	7	6	9	2	5	8
5	9	6	4	8	2	1	7	3

EXPERT - 456

9	5	4	2	3	1	6	8	7
3	6	7	5	9	8	4	1	2
1	2	8	6	7	4	5	3	9
2	1	9	3	6	5	7	4	8
7	8	6	4	1	9	3	2	5
5	4	3	8	2	7	1	9	6
4	7	1	9	8	6	2	5	3
8	3	5	7	4	2	9	6	1
6	9	2	1	5	3	8	7	4

EXPERT - 457

6	2	4	5	1	3	8	9	7
9	8	5	7	4	6	3	1	2
3	1	7	8	2	9	6	5	4
7	3	1	6	5	4	2	8	9
8	9	6	2	7	1	5	4	3
4	5	2	3	9	8	1	7	6
5	7	8	9	6	2	4	3	1
1	6	3	4	8	7	9	2	5
2	4	9	1	3	5	7	6	8

EXPERT - 458

5	8	4	9	7	3	1	2	6
6	1	3	5	8	2	7	9	4
7	9	2	1	6	4	3	8	5
3	2	8	4	9	7	5	6	1
1	5	6	3	2	8	9	4	7
4	7	9	6	5	1	2	3	8
2	6	5	8	1	9	4	7	3
8	3	7	2	4	5	6	1	9
9	4	1	7	3	6	8	5	2

EXPERT - 459

6	4	5	2	7	3	8	1	9
3	7	9	4	1	8	6	5	2
1	8	2	6	5	9	3	7	4
5	2	7	3	4	6	1	9	8
8	6	4	5	9	1	7	2	3
9	3	1	8	2	7	5	4	6
7	1	6	9	8	4	2	3	5
4	5	8	7	3	2	9	6	1
2	9	3	1	6	5	4	8	7

EXPERT - 460

1	7	8	5	6	3	9	2	4
5	9	6	1	4	2	7	3	8
4	2	3	9	8	7	1	5	6
8	3	7	2	1	6	4	9	5
6	1	2	4	5	9	3	8	7
9	4	5	3	7	8	2	6	1
2	6	1	7	3	5	8	4	9
3	8	4	6	9	1	5	7	2
7	5	9	8	2	4	6	1	3

EXPERT - 461

9	6	1	7	5	3	2	4	8
8	7	4	1	9	2	3	6	5
2	3	5	4	8	6	7	1	9
7	4	9	5	6	1	8	2	3
3	2	8	9	7	4	6	5	1
1	5	6	3	2	8	4	9	7
4	8	7	2	1	5	9	3	6
5	9	3	6	4	7	1	8	2
6	1	2	8	3	9	5	7	4

EXPERT - 462

3	5	4	8	1	7	9	6	2
8	2	1	6	9	5	4	7	3
7	9	6	3	2	4	5	1	8
9	6	8	5	7	2	3	4	1
2	1	3	4	8	6	7	5	9
5	4	7	9	3	1	8	2	6
1	8	2	7	4	9	6	3	5
4	3	5	2	6	8	1	9	7
6	7	9	1	5	3	2	8	4

EXPERT - 463

7	5	6	1	4	2	3	8	9
4	8	1	6	3	9	7	5	2
3	2	9	5	7	8	4	1	6
1	3	2	9	6	5	8	7	4
9	4	5	8	1	7	2	6	3
6	7	8	3	2	4	1	9	5
2	6	7	4	5	1	9	3	8
8	1	3	2	9	6	5	4	7
5	9	4	7	8	3	6	2	1

EXPERT - 464

9	8	4	2	6	3	1	7	5
5	3	1	4	7	8	6	9	2
6	7	2	5	9	1	8	3	4
2	5	3	1	4	7	9	6	8
4	1	9	6	8	5	7	2	3
8	6	7	9	3	2	5	4	1
1	4	6	8	2	9	3	5	7
3	2	5	7	1	6	4	8	9
7	9	8	3	5	4	2	1	6

EXPERT - 465

8	6	1	4	3	2	5	9	7
9	4	3	6	5	7	8	1	2
7	2	5	9	1	8	4	3	6
6	1	2	7	4	3	9	8	5
5	9	7	2	8	1	6	4	3
4	3	8	5	9	6	2	7	1
3	7	6	8	2	9	1	5	4
1	8	4	3	6	5	7	2	9
2	5	9	1	7	4	3	6	8

EXPERT - 466

6	8	5	9	4	7	1	2	3
4	9	7	2	3	1	6	5	8
3	2	1	6	8	5	7	4	9
8	1	4	3	7	2	9	6	5
5	3	9	8	1	6	2	7	4
2	7	6	5	9	4	3	8	1
9	4	3	7	6	8	5	1	2
1	6	2	4	5	3	8	9	7
7	5	8	1	2	9	4	3	6

EXPERT - 467

3	2	8	9	4	1	5	7	6
5	6	9	3	7	2	4	1	8
4	1	7	6	8	5	3	2	9
8	7	3	4	1	9	2	6	5
9	4	2	5	6	7	1	8	3
6	5	1	2	3	8	7	9	4
7	3	4	8	2	6	9	5	1
2	9	6	1	5	4	8	3	7
1	8	5	7	9	3	6	4	2

EXPERT - 468

6	2	4	3	8	7	9	1	5
7	8	9	1	5	6	2	3	4
3	5	1	2	4	9	7	6	8
1	3	7	9	2	5	8	4	6
5	9	8	6	7	4	1	2	3
2	4	6	8	3	1	5	9	7
4	1	3	7	9	8	6	5	2
8	6	2	5	1	3	4	7	9
9	7	5	4	6	2	3	8	1

EXPERT - 469

2	6	8	3	1	7	5	9	4
1	7	4	6	9	5	2	8	3
3	5	9	8	4	2	6	1	7
7	9	3	1	2	8	4	5	6
6	1	2	4	5	3	9	7	8
8	4	5	9	7	6	1	3	2
5	8	6	2	3	1	7	4	9
9	3	7	5	6	4	8	2	1
4	2	1	7	8	9	3	6	5

EXPERT - 470

1	2	4	3	7	5	8	9	6
5	6	8	2	9	4	7	3	1
3	9	7	8	1	6	4	5	2
6	4	2	7	3	9	5	1	8
8	1	9	5	4	2	6	7	3
7	5	3	6	8	1	2	4	9
2	3	6	1	5	7	9	8	4
4	8	5	9	2	3	1	6	7
9	7	1	4	6	8	3	2	5

EXPERT - 471

8	7	1	4	6	9	3	2	5
2	5	3	7	8	1	9	6	4
4	9	6	2	3	5	1	8	7
9	6	5	8	2	4	7	3	1
7	8	4	1	5	3	2	9	6
1	3	2	9	7	6	5	4	8
3	1	7	6	9	8	4	5	2
5	4	8	3	1	2	6	7	9
6	2	9	5	4	7	8	1	3

EXPERT - 472

9	4	2	5	7	1	8	6	3
6	5	8	9	3	2	7	4	1
1	7	3	4	8	6	5	9	2
5	6	1	7	2	9	4	3	8
8	3	4	6	1	5	9	2	7
7	2	9	3	4	8	1	5	6
3	9	5	1	6	7	2	8	4
4	8	7	2	5	3	6	1	9
2	1	6	8	9	4	3	7	5

EXPERT - 473

9	3	1	5	2	4	6	8	7
6	5	8	7	1	3	9	2	4
7	2	4	9	6	8	5	3	1
5	1	7	8	4	2	3	6	9
2	8	9	3	7	6	4	1	5
4	6	3	1	9	5	2	7	8
1	7	6	4	3	9	8	5	2
8	9	2	6	5	1	7	4	3
3	4	5	2	8	7	1	9	6

EXPERT - 474

2	3	5	4	7	6	9	1	8
6	4	8	2	9	1	3	7	5
7	1	9	3	8	5	2	6	4
3	8	1	7	6	2	4	5	9
4	6	7	5	3	9	1	8	2
5	9	2	8	1	4	6	3	7
1	2	3	9	5	7	8	4	6
8	7	4	6	2	3	5	9	1
9	5	6	1	4	8	7	2	3

EXPERT - 475

2	4	9	7	6	8	3	1	5
7	3	8	1	5	2	6	4	9
6	5	1	4	9	3	2	8	7
8	6	3	5	4	9	1	7	2
1	9	5	3	2	7	8	6	4
4	7	2	8	1	6	9	5	3
9	8	6	2	7	5	4	3	1
5	2	4	6	3	1	7	9	8
3	1	7	9	8	4	5	2	6

EXPERT - 476

3	1	5	9	8	2	4	6	7
6	7	2	1	4	3	9	8	5
9	8	4	6	7	5	2	3	1
1	9	7	3	5	6	8	4	2
2	4	8	7	9	1	6	5	3
5	3	6	8	2	4	1	7	9
7	2	3	4	1	8	5	9	6
4	6	1	5	3	9	7	2	8
8	5	9	2	6	7	3	1	4

EXPERT - 477

9	4	6	5	1	7	2	8	3
2	8	3	4	9	6	5	7	1
7	1	5	2	3	8	9	6	4
4	2	1	6	8	5	7	3	9
5	7	9	1	4	3	8	2	6
3	6	8	7	2	9	4	1	5
1	3	4	9	7	2	6	5	8
6	9	2	8	5	1	3	4	7
8	5	7	3	6	4	1	9	2

EXPERT - 478

2	6	1	8	7	9	3	4	5
5	7	4	6	1	3	8	2	9
3	9	8	4	5	2	1	6	7
1	5	3	7	8	6	4	9	2
6	2	9	1	3	4	5	7	8
8	4	7	2	9	5	6	3	1
9	8	5	3	6	7	2	1	4
7	3	2	5	4	1	9	8	6
4	1	6	9	2	8	7	5	3

EXPERT - 479

7	2	1	4	8	3	5	6	9
5	6	8	1	7	9	4	2	3
4	3	9	6	2	5	7	1	8
6	8	7	5	9	2	3	4	1
1	5	3	7	4	6	8	9	2
9	4	2	8	3	1	6	7	5
3	1	5	9	6	4	2	8	7
2	7	4	3	1	8	9	5	6
8	9	6	2	5	7	1	3	4

EXPERT - 480

9	7	6	8	1	3	4	5	2
2	8	1	9	4	5	7	6	3
5	3	4	2	7	6	9	8	1
7	2	5	4	8	9	3	1	6
1	4	3	6	5	7	2	9	8
8	6	9	3	2	1	5	7	4
6	9	7	1	3	2	8	4	5
4	5	2	7	6	8	1	3	9
3	1	8	5	9	4	6	2	7

EXPERT - 481

3	4	2	9	5	6	7	8	1
9	5	7	1	8	3	4	6	2
6	8	1	7	4	2	3	5	9
1	2	9	8	7	5	6	4	3
8	7	3	2	6	4	9	1	5
5	6	4	3	9	1	8	2	7
7	9	5	4	2	8	1	3	6
4	3	6	5	1	7	2	9	8
2	1	8	6	3	9	5	7	4

EXPERT - 482

1	7	2	6	4	3	8	5	9
3	9	6	2	8	5	4	1	7
4	5	8	7	1	9	3	2	6
9	2	7	4	5	6	1	8	3
6	3	4	8	9	1	5	7	2
8	1	5	3	7	2	9	6	4
5	6	1	9	3	7	2	4	8
2	8	9	5	6	4	7	3	1
7	4	3	1	2	8	6	9	5

EXPERT - 483

7	3	1	5	6	9	4	2	8
2	6	8	7	1	4	5	9	3
9	4	5	8	3	2	7	6	1
3	8	4	9	5	6	1	7	2
1	9	6	3	2	7	8	4	5
5	2	7	1	4	8	9	3	6
6	5	9	2	7	1	3	8	4
8	1	2	4	9	3	6	5	7
4	7	3	6	8	5	2	1	9

EXPERT - 484

7	2	3	9	4	6	5	1	8
4	5	9	1	8	7	2	3	6
1	6	8	3	2	5	9	7	4
8	9	2	7	1	3	6	4	5
5	1	7	4	6	9	3	8	2
6	3	4	8	5	2	7	9	1
2	7	1	6	9	4	8	5	3
3	8	5	2	7	1	4	6	9
9	4	6	5	3	8	1	2	7

EXPERT - 485

6	8	9	1	4	7	5	2	3
3	2	1	5	6	9	7	4	8
5	4	7	2	8	3	1	6	9
9	3	4	7	2	6	8	1	5
7	6	8	3	1	5	4	9	2
2	1	5	4	9	8	6	3	7
8	9	3	6	7	4	2	5	1
1	7	6	9	5	2	3	8	4
4	5	2	8	3	1	9	7	6

EXPERT - 486

1	5	9	4	2	6	7	3	8
6	4	7	9	3	8	1	2	5
3	8	2	1	5	7	4	9	6
5	3	1	7	9	2	8	6	4
7	6	8	3	4	5	2	1	9
9	2	4	6	8	1	3	5	7
4	7	3	5	1	9	6	8	2
8	9	6	2	7	3	5	4	1
2	1	5	8	6	4	9	7	3

EXPERT - 487

2	8	1	3	9	5	6	4	7
4	3	5	1	6	7	2	8	9
9	6	7	4	8	2	5	1	3
5	9	4	2	3	8	7	6	1
3	1	8	9	7	6	4	2	5
6	7	2	5	4	1	3	9	8
7	2	3	8	1	4	9	5	6
8	4	9	6	5	3	1	7	2
1	5	6	7	2	9	8	3	4

EXPERT - 488

5	4	8	9	1	2	6	3	7
2	7	3	4	8	6	5	9	1
9	1	6	7	3	5	4	2	8
7	3	9	6	4	8	2	1	5
8	5	4	1	2	3	9	7	6
6	2	1	5	9	7	8	4	3
3	8	7	2	5	9	1	6	4
1	9	5	3	6	4	7	8	2
4	6	2	8	7	1	3	5	9

EXPERT - 489

7	9	6	2	5	1	8	4	3
2	3	4	8	7	6	1	5	9
5	1	8	3	9	4	7	6	2
8	5	2	6	3	7	4	9	1
3	6	1	9	4	2	5	7	8
9	4	7	5	1	8	2	3	6
6	7	3	1	8	5	9	2	4
1	2	5	4	6	9	3	8	7
4	8	9	7	2	3	6	1	5

EXPERT - 490

4	3	7	9	5	2	1	8	6
8	1	5	6	4	7	2	9	3
9	2	6	3	1	8	5	4	7
5	8	1	4	6	9	7	3	2
2	9	3	7	8	5	6	1	4
7	6	4	1	2	3	9	5	8
3	5	2	8	9	6	4	7	1
6	4	8	5	7	1	3	2	9
1	7	9	2	3	4	8	6	5

EXPERT - 491

3	9	6	5	1	4	7	2	8
7	4	5	3	2	8	1	9	6
1	8	2	6	9	7	4	3	5
2	7	8	9	3	1	5	6	4
4	6	3	8	5	2	9	7	1
5	1	9	7	4	6	3	8	2
8	2	1	4	7	9	6	5	3
6	3	7	1	8	5	2	4	9
9	5	4	2	6	3	8	1	7

EXPERT - 492

4	7	3	2	1	8	5	6	9
6	5	1	4	7	9	2	8	3
8	2	9	3	6	5	1	4	7
3	6	5	9	8	4	7	1	2
9	4	7	1	3	2	8	5	6
1	8	2	7	5	6	9	3	4
5	9	8	6	4	7	3	2	1
2	1	4	5	9	3	6	7	8
7	3	6	8	2	1	4	9	5

EXPERT - 493

3	5	8	2	6	7	9	4	1
2	6	4	9	1	3	7	8	5
7	9	1	5	8	4	6	2	3
9	4	5	7	2	1	3	6	8
1	2	3	8	9	6	5	7	4
6	8	7	4	3	5	1	9	2
5	7	9	3	4	8	2	1	6
8	1	2	6	5	9	4	3	7
4	3	6	1	7	2	8	5	9

EXPERT - 494

9	4	7	3	5	8	6	2	1
5	6	2	9	4	1	7	8	3
8	3	1	6	2	7	4	9	5
4	2	5	7	6	3	9	1	8
7	1	8	4	9	5	2	3	6
3	9	6	1	8	2	5	7	4
6	7	4	8	3	9	1	5	2
1	5	3	2	7	4	8	6	9
2	8	9	5	1	6	3	4	7

EXPERT - 495

7	3	2	8	1	4	6	9	5
4	5	8	3	9	6	2	1	7
1	6	9	5	2	7	8	3	4
6	8	7	1	4	5	3	2	9
5	2	1	6	3	9	4	7	8
9	4	3	7	8	2	5	6	1
8	7	4	2	6	1	9	5	3
3	1	6	9	5	8	7	4	2
2	9	5	4	7	3	1	8	6

EXPERT - 496

2	8	3	4	5	9	1	7	6
7	1	9	8	2	6	4	5	3
6	4	5	3	7	1	8	2	9
1	3	7	5	8	2	9	6	4
9	6	2	7	1	4	3	8	5
4	5	8	6	9	3	7	1	2
5	2	4	1	3	8	6	9	7
8	9	6	2	4	7	5	3	1
3	7	1	9	6	5	2	4	8

EXPERT - 497

1	2	4	3	8	7	6	9	5
8	5	9	4	6	1	7	2	3
7	3	6	9	5	2	8	4	1
6	7	8	2	4	5	3	1	9
3	1	5	6	7	9	4	8	2
4	9	2	1	3	8	5	6	7
9	8	3	5	1	4	2	7	6
2	6	7	8	9	3	1	5	4
5	4	1	7	2	6	9	3	8

EXPERT - 498

8	9	6	7	1	3	2	5	4
1	4	3	2	9	5	8	6	7
2	5	7	8	6	4	3	9	1
6	3	8	5	4	9	7	1	2
9	7	2	6	8	1	5	4	3
5	1	4	3	2	7	6	8	9
4	6	5	1	3	2	9	7	8
3	8	9	4	7	6	1	2	5
7	2	1	9	5	8	4	3	6

EXPERT - 499

9	8	6	5	1	2	3	7	4
7	2	5	3	9	4	8	6	1
3	4	1	7	8	6	9	2	5
6	3	8	1	7	9	4	5	2
4	7	2	6	5	8	1	3	9
5	1	9	4	2	3	7	8	6
1	9	7	8	6	5	2	4	3
8	6	4	2	3	1	5	9	7
2	5	3	9	4	7	6	1	8

EXPERT - 500

3	7	4	2	1	9	8	5	6
9	8	5	7	4	6	1	3	2
2	6	1	8	3	5	9	7	4
8	3	9	5	6	4	2	1	7
7	5	6	1	9	2	3	4	8
1	4	2	3	8	7	6	9	5
4	2	8	9	5	1	7	6	3
5	1	7	6	2	3	4	8	9
6	9	3	4	7	8	5	2	1